THE MATHEMATICS OF FINANCE FOR HIGH SCHOOLERS

First published 2023 by Panda Ohana Publishing

pandaohana.com

admin@pandaohana.com

First printing 2023

©2023 Paul Wilmott

All rights reserved. This book or any portion thereof may not be reprinted or reproduced or utilized in any form or by any electronic, mechanical or other means, now known or hereafter invented, including photocopying, scanning or recording, or in any information and retrieval system, without the express written permission of the publisher. The publisher and the author of this book shall not be liable in the event of incidental or consequential damages in connection with, or arising out of reference to or reliance on any information in this book or use of the suggestions contained in any part of this book. The publisher and the author make no representation or warranties with respect to the accuracy or completeness of the contents of this book and specifically disclaim all warranties, including without limitation warranties of fitness for a particular purpose. The methods contained herein may not be suitable for every situation. The information in this book does not constitute advice.

ISBN 978-1-9160816-5-9

THE MATHEMATICS OF FINANCE FOR HIGH SCHOOLERS

Paul Wilmott

Panda Ohana Publishing

Other books by Paul Wilmott

Option Pricing: Mathematical Models and Computation. (With J.N.Dewynne and S.D.Howison.) Oxford Financial Press (1993)

Mathematics of Financial Derivatives: A Student Introduction. (With J.N.Dewynne and S.D.Howison.) Cambridge University Press (1995)

Mathematical Models in Finance. (Ed. with S.D.Howison and F.P.Kelly.) Chapman and Hall (1995)

Derivatives: The Theory and Practice of Financial Engineering. John Wiley and Sons (1998)

Paul Wilmott On Quantitative Finance. John Wiley and Sons (2000)

Paul Wilmott Introduces Quantitative Finance. John Wiley and Sons (2001)

New Directions in Mathematical Finance. (Ed. with H.Rasmussen.) John Wiley and Sons (2001)

Best of Wilmott 1. (Ed.) John Wiley and Sons (2004)

Exotic Option Pricing and Advanced Levy Models. (Ed. with W.Schoutens and A.Kyprianou) John Wiley and Sons (2005)

Best of Wilmott 2. (Ed.) John Wiley and Sons (2005)

Frequently Asked Questions in Quantitative Finance. John Wiley and Sons (2006)

Paul Wilmott On Quantitative Finance, second edition. John Wiley and Sons (2006)

Paul Wilmott Introduces Quantitative Finance, second edition. John Wiley and Sons (2007)

Frequently Asked Questions in Quantitative Finance, second edition. John Wiley and Sons (2009)

The Money Formula: Dodgy Finance, Pseudo Science, and How Mathematicians Took Over the Markets. (With D.Orrell). John Wiley and Sons (2017)

Machine Learning: An applied mathematics introduction. Panda Ohana Publishing (2019)

The Mathematics of Artificial Intelligence for High Schoolers. Panda Ohana Publishing (2023)

For all of my students

'I love money. I love everything about it. I bought some pretty good stuff. Got me a $300 pair of socks. Got a fur sink. An electric dog polisher. A gasoline powered turtleneck sweater. And, of course, I bought some dumb stuff, too.' Steve Martin

Contents

PREFACE	ix
INTRODUCTION	1
Definitions	1
A Cheap History Of Money	2
Banks	3
Bank Runs, Bubbles And Crashes	4
Important Financial Models And Theories	8
SOME JARGON	11
CAREERS IN FINANCE	25
Non-technical Careers	25
Technical Careers	27
SOME MATHEMATICS	30
Expectation	30
Standard Deviation	32
Correlation	34
IN DEPTH: FIXED INCOME	37
The Mathematics	38
What Next?	47
Exercises	48
IN DEPTH: SHARES	50
The Mathematics	51
What Next?	62
Exercises	63

IN DEPTH: PORTFOLIOS	65
The Mathematics	65
What Next?	73
Exercises	74
IN DEPTH: OPTIONS	75
The Mathematics	77
What Next?	94
Exercises	95
APPENDIX	96
Exponential Function	96
Logarithms	98
Continuous Compounding	100
Annual Rate Versus Continuous Compounding	101
Floating Interest Rates	102
Probability Distributions	104
The Law Of Large Numbers	106
The Central Limit Theorem	107
The Normal Distribution	112
Normally Distributed Share Price Returns	113
The Distribution Of Share Prices	114
Option Replication	117
EPILOGUE	119
ACKNOWLEDGMENTS	119
INDEX	120
ABOUT PAUL WILMOTT	124
ALSO BY PAUL WILMOTT	125

PREFACE

'Money makes the world go around' That's according to Fred Ebb. He went on to expand, 'The world go around, The world go around, Money makes the world go around, It makes the world go 'round.' His lyrics to the music of John Kandor, sung by Liza Minnelli and Joel Grey in the film version of *Cabaret*, direction and choreography by Bob Fosse, were pretty emphatic. More details were supplied by Benny Andersson and Björn Ulvaeus, and sung by Anni-Frid Lyngstad, and Agnetha Fältskog, 'All the things I could do, If I had a little money, It's a rich man's world.'

True.

So now I'm wondering why you are reading this book. Were you drawn in by the word 'Mathematics' in the title? It could be that you love to learn about the real-world applications of your favourite subject. Or maybe it was the 'Finance' that grabbed you, perhaps you are thinking far ahead to a well-compensated career in banking. Both are perfectly acceptable motives. Yes, finance is a great source for interesting mathematical problems, some hinted at in this book. And yes, finance can be a very highly paid employment choice, more details to follow. Either way I hope you find this book informative.

Before I learned about the mathematics of finance some [deleted] years ago, I thought that finance as a career was inhabited by privately educated, middle-aged gentlemen in chalk-striped suits, doing deals in smoke-filled clubs in and around St James's in London. Wearing a bowler hat. Or twenty-somethings who left school at 16 with no qualifications, with a 'phone at each ear, shouting abuse across a paper-strewn trading floor. Also smoke filled. But no hats. Well, it's not like that anymore. And hasn't

> '**Most people work just hard enough not to get fired and get paid just enough money not to quit.**' George Carlin

been for quite a while. A lot of people working in finance these days have studied mathematics at university, many up to master's or even doctorate level. And I know a lot of mathematicians, none of whom wear chalk-striped suits or smoke. And only a few of them shout abuse. Many mathematicians working in finance are writing computer programmes to implement theories to predict how financial markets behave, increasingly using artificial intelligence. Some mathematicians are even developing those theories, in universities and banks. Finance now is about statistics, probabilities, calculus, machine learning, and so on, all very technical *mathematical* subjects.

Of course, you don't need to know about these subjects to work in many, even most, finance jobs. Familiarity with a few basic concepts and a decent competency with a spreadsheet will be enough. But since you are reading this book, I'm going to assume that you are curious about the role of mathematics in finance. How much mathematics is there in finance, is it interesting mathematics, will you be able to cope? The answers are, in order, potentially lots, yes it can be, and YES!

In this book you'll read about the roots of money, how Fred Ebb was correct that it makes the world go around. You'll see a few mathematical theories. You'll also see why some of these theories don't make much sense. You'll see smart people who won Nobel Prizes for their work. And you'll see how some of these same smart people lost lots of money investing. Oh, yes, it's a fun subject!

To get you in the right mood there's one thing I want to mention early on. This is something that my own research has focused on to some extent, and which is probably a bit unusual in any book with mathematics in the title. It stems from the above sentence, 'You'll also see why some of these theories don't make much sense.' If this was a book about the mathematics of geometry, or physics, or biochemistry, say, then I might give you some history about theories that turned out to be wrong, and got replaced with better theories, the old theories never to be used again. Finance is strangely not like that. There are theories out there that can easily be shown to be nonsense but are still being used every day. How can a theory that's blatantly wrong still be used in something as important as finance? Let me justify this, kind of, by

> **'Money will buy a pretty good dog, but it won't buy the wag of his tail.'** Josh Billings

saying the following: It might not matter. Too much. One possible justification is that, ok, the theories don't work at the small scale of buying and selling individual financial products but when scaled up to a few quadrillion dollars somehow the details don't matter, that some law of averages kicks in. Also part of this is to say, well, if we didn't have any theories at all, even the bad ones, then no one would ever trade anything. And where would we be if there was no trading? I'll tell you where, still living in caves, bartering with turnips. There's another justification, somewhat cynical, but also probably close to true, that there's sooooooo much money sloshing around in banking that again the niceties of a theory don't matter too much. And you will know from personal experience that unless you pay for everything in cash that some bank somewhere is taking a percentage, their cut, of every transaction you make. And you don't need a theory for that!

If you think it's bad to use theories that are wrong then I suggest you never, ever study economics. That's even worse!

[Editor: 'Enough of the negativity, Paul.']

In this book we'll be looking at many aspects of finance, from its history, the jargon, through careers, and then, of course, there'll be plenty of mathematics. The mathematics covered here is not the mathematics of balancing your check book, or splitting a restaurant bill among friends, some of whom had the set menu, and others the specials. The mathematics here is an introduction to the subject known as mathematical finance, quantitative finance or financial engineering. It's an introduction to the higher level of mathematics used in banks and hedge funds, in valuing complex financial instruments, in allocating money to different assets, and in measuring and managing risk.

Not all jobs in finance require high-level mathematics, but if you do enjoy mathematics at a high level then there are jobs in finance that require it.

At the end of later chapters there are some simple exercises for you to try, to help cement the ideas.

On the subject of exercises, if you want me to send you the answers then you need to do something for me first ... email me your answers to the first *two* questions in the Exercise sections of each of the last four chapters! I'm paul@wilmott.com. Then you'll get the answers to *all* questions in return. That's assuming your answers show at least some plausible attempt, I mean you can't just say '42.' I should have been a lawyer.

Also, email me if you want some basic spreadsheets that illustrate the mathematical ideas. There's one spreadsheet for each of the last five chapters and one for the appendix.

Yes, there is an appendix near the end of the book. This is for the mathematical material that you might want to skip. At least on a first reading. This is material that you might have seen in high school, but then again you might not. It's not necessary for understanding most of the basics of financial mathematics, but if you study the subject at university it will become crucial. The Appendix mostly involves the special functions logarithms and exponentials, and extra material on probability theory.

INTRODUCTION

Definitions

We start with a definition from Wikipedia:

Finance is the study and discipline of money, currency and capital assets. It is related to, but not synonymous with economics, which is the study of production, distribution, and consumption of money, assets, goods and services (the discipline of financial economics bridges the two). Finance activities take place in financial systems at various scopes; thus, the field can be roughly divided into personal, corporate, and public finance.

We shall be avoiding economics as much as humanly possible. (I have views on the subject of economics which might inadvertently creep into this book!)

Further:

In a financial system, assets are bought, sold, or traded as financial instruments, such as currencies, loans, bonds, shares, stocks, options, futures, etc. Assets can also be banked, invested, and insured to maximize value and minimize loss.
In practice, risks are always present in any financial action and entities.

> 'I've got all the money I'll ever need, if I die by four o'clock.'
> Henny Youngman

There is a lot of jargon in finance, I'll have to define quite a lot of technical terms, a whole chapter's worth. And there are a lot of different financial products, with more being created all the time. Although I'll keep things as simple as possible. I've written other, very technical, books in this field if you are feeling strong, but TBH you should avoid those for now ... enjoy your schooldays while you can!

If it all gets too much, just repeat, 'It's only money.'

A Cheap History Of Money

(Apologies to Stephen Hawking, but I couldn't resist.)

> **'Ben Franklin may have discovered electricity, but it is the man who invented the meter who made the money.'** Earl Warren

Before there was money there was bartering. Bartering is simply the exchange of one person's goods or services for another person's goods or services. Person A has more chickens than they need, Person B has too many potatoes. Person A values a few potatoes more than they value one of their chickens and Person B values them the other way around. So they swap. (Millenia would pass, and the development of plastics, before the first Happy Meal was created.)

It had been assumed, by people who study these things, that this exchange happened simultaneously, but now it is thought that the mechanism for exchange was more likely to be via one-way gifts, a buildup of trust, and then, later, a gift in return. Otherwise, there would be what is known as the 'double coincidence of wants,' you've got the chickens but the person with the potatoes and an urgent need for a chicken is nowhere to be found. This gift economy may date back 100,000 years. After barter came commodity money. This is a form of money linked to the worth of whatever it is made from, such as grain. This was around 3,000 years ago. Slightly more abstract, in the sense that its value is in the eye of the beholder, was money consisting of shells, used commonly on many continents.

And then around 600 BC, gold and silver coins were first created, 'minted,' in what is now western Turkey. Merchants who dealt in gold and silver would issue paper receipts to customers representing how much gold or silver they had bought. These receipts were then used for payment, an early form of paper money.

The first genuine paper money dates back to 11th century China. But the concept didn't reach Europe until the 13th century, and that was thanks to travellers such as Marco Polo.

Sometimes the worth of a currency is linked to something physical that is inherently valuable, such as gold again. In the early 20th century one US dollar was fixed at 1.50463g of gold. This link between the US dollar and gold ended in the early 1970s. Sometimes money has nothing behind it, a government issues money that has no support other than faith in the issuing government, that's called *fiat* money, from the Latin meaning 'let it be done.'

> **'Business is the art of extracting money from another man's pocket without resorting to violence.'** Max Amsterdam

And now there's cryptocurrency! A cryptocurrency, or simply crypto, is a medium of exchange like currencies but one that does not depend on any bank, does not come in any physical form, and ownership records are stored in a computerized database. Advantages of cryptos are in cheaper transactions and in the decentralization that supposedly reduces risk of collapse. On the other hand, the sudden extreme rises and falls in the value of cryptos relative to other currencies is an obvious downside. As is the use of cryptos in criminal activities thanks to their anonymity. (As I write this, I have just received a spam email containing a supposed analysis of crypto markets. Such obvious spam doesn't help in making crypto respectable!) Several winners of the Nobel Prize in Economics have called crypto a bubble.

Banks

Banks date back to mediaeval and Renaissance Italy, created by families such as the Medici. There are several different types of bank. The basic retail bank that you will be most familiar with is an institution that takes in money from people who want to save it for a rainy day and then lend it out to people who need more money than they currently have so they can buy expensive items such as a house. Other banks, investment banks, only deal with companies as clients and help them with complex deals such as raising cash for expansion or arranging mergers and acquisitions. There are also Central Banks, usually associated with a country. These institutions regulate other banks, to prevent their collapse, and they control policies involving money. Examples are the Bank of England, for the United Kingdom not just the English bit, on which other central banks were based, the European Central Bank, for the European Union and the Federal Reserve in the United States.

Bank Runs, Bubbles And Crashes

Bank runs, bubbles and crashes are financial phenomena classically unrelated to actual values of assets but are often psychological in nature, sometimes caused by human herding behaviour.

Keeping your money in a bank requires trust in that bank. If your bank is investing in assets that are illiquid, meaning they can't be easily turned into cash, in such amounts that the bank itself might fail if/when those assets fall in value, then the natural reaction is to take your money out of the bank asap, and put it somewhere safer. 'Asap' means before too many other people get the same idea. Once many people start to take out their money, panic follows, and the result is a run on the bank.

The Panic of 1907 was sparked by the collapse of the Knickerbocker Trust Company. This led to a snowballing of bank runs throughout the United States. People lost confidence in banks, and lines formed outside institutions as depositors queued to withdraw their savings. JP Morgan organized a group of bankers to provide financial support to troubled banks. This panic demonstrated the need for a central bank, and this led to the creation of the Federal Reserve System in the United States.

An early bubble and crash combo was the infamous Tulip Mania of 1634-7. Tulip bulbs in the Netherlands became so fashionable that over a very short period of time the price of a single bulb reached that of a house. And then, even faster, the prices plummeted. So extreme was this particular bubble-then-crash that it has become the go-to reference for anything else that looks like a bubble, often warning of a following crash. A more recent example was the dot-com bubble of the late 1990s followed by the crash of late 2002. Investors put their money in anything technology related, common sense had left the building.

> 'October. This is one of the peculiarly dangerous months to speculate in stocks. The others are July, January, September, April, November, May, March, June, December, August, and February.' *Pudd'nhead Wilson* by Mark Twain

Explanations for bubbles and crashes are the possibly irrational herding, mentioned above, and the rational but risky Greater Fool Theory, that underlying value is irrelevant as long as you can find someone (more daft than yourself) to buy from you at an even higher price.

House prices are often thought to be in a bubble. But houses prices can be linked to income and interest rates, whereas 'a short stem with fleshy leaves or leaf bases that function as food storage organs during dormancy,' [Wikipedia] is harder to value, as is new technology. Having said that, houses played an important part in the credit crisis of 2007-8. Banks would lend to people wanting to buy a house with little concern for their ability to pay off the loan.

Banks would then package up these loans to pass on to other investors (see the Greater Fool Theory above!). When house prices collapsed it caused a global financial crisis, with major banks and financial institutions facing insolvency, and then a severe global recession.

To give you a sense of what it's like during a crash, here is a story from the *Los Angeles Times* of October 20th, 1987:

> **BEDLAM ON WALL STREET : Stocks Plunge 508 Amid Panic : Record One-Day Decline Eclipses 1929 Market Drop : Huge Losses Raise Fears for Economy BY BILL SING**
>
> *Fear and panic of historic proportions overwhelmed the stock market Monday, forcing the Dow Jones industrial average to a stunning decline of 508 points, by far the biggest one-day plunge in history.*
>
> *The Dow Jones average lost almost a quarter of its value, nearly doubling the percentage drop that occurred on Oct. 28, 1929, the crash that preceded the Great Depression. With Monday's fall, the Dow has lost half the ground it had gained since the bull market got under way on Aug. 12, 1982, when the Dow stood at a mere 776.92.*
>
> *The collapse came on dizzying trading volume of 604.4 million shares on the New York Stock Exchange, nearly double the previous record of 338.5 million shares, set last Friday.*
>
> *The unprecedented dive in the Dow, to a close of 1,738.74, immediately raised fears of an international economic crisis and a recession in the United States. Mutual funds, pension funds, foundations, endowment funds, retirement accounts, stock brokerage firms, individual investors and others together lost hundreds of billions of dollars, and the impact could reverberate throughout the U.S. economy.*
>
> *Monday's debacle alone wiped out $503 billion in the market value of all stocks. A total of about $1 trillion in stock market wealth has been erased since the market peaked on Aug. 25, as measured by Wilshire Associates of Los Angeles.*
>
> *…*
>
> *The collapse spurred calls among Wall Street and congressional officials for dramatic government action, ranging from pumping more*

> *money into battered financial markets to regulation of certain market practices that may have aggravated Monday's free fall.*
>
> *...*
>
> *'It was a financial meltdown as good as any that has come along,' said John Phelan, chairman of the New York Stock Exchange, in describing Monday's rout as the financial equivalent of the Chernobyl nuclear disaster in the Soviet Union last year.*
>
> *...*
>
> *Some analysts said it could take weeks, if not months, before investors regain their confidence in the market.*
>
> *Others, however, said a turnaround could be just around the corner, and that many bargains can be found.*

You've gotta love the optimism in that last sentence. (We'll see a financial version of hedging soon!) But it did only take two years for the Dow Jones index to recover. For comparison, after the 1929 crash it took 25 years before the index returned to its pre-crash level.

I'd like to end this section with a mention of Long Term Capital Management (LTCM). This financial firm nicely links the financial crises above with the financial theories coming next. LTCM was founded by prominent people in finance, including two winners of the 1997 Nobel Memorial Prize in Economic Sciences, Professors Robert C. Merton and Myron Scholes (pronounced like shoals). They built their business on mathematical modelling of various financial instruments, exploiting what they perceived as mispricings. Unfortunately, 1998 saw major financial crises in Russia and elsewhere, and LTCM's elegant mathematical theories broke down. LTCM was bailed out by major financial institutions, and this prevented a broader financial meltdown.

> **'If you have trouble imagining a 20% loss in the stock market, you shouldn't be in stocks.'** John Bogle

Important Financial Models And Theories

Finance is jam packed with models and theories. Some are quite good, some not so much!

Bachelier's model of share prices

In the late 1890s as a doctoral student Louis Bachelier developed a mathematical model for the behaviour of stock prices and presented some early ideas around the valuation of the clever financial instruments known as options, of which a lot more later. His model for share prices treated them as moving randomly. Whether a stock moved up or down one day and by how much was decided by, mathematically speaking, pulling a random number out of a hat. And then we pull out another number to represent the share move the day after, and so on.

Along the way Bachelier pretty much invented the mathematical subject of stochastic calculus. He showed how this mathematics also modelled diffusion, such as in the diffusion of smoke in a room or coloured dye in a swimming pool. This work on diffusion predated Einstein's by five years, and his basic assumptions were essentially those of the Efficient Market Hypothesis below.

Another interesting part of his story is that, thanks to the snobbishness of academics towards money, eurgh, his work was then ignored, except by some hard-core mathematicians, for many decades. However, now he is recognized as the father of financial mathematics.

Modern Portfolio Theory

Modern Portfolio Theory (MPT) is one of the foundations of finance, developed by Harry Markowitz in the 1950s. Again, he assumed that the prices of shares move at random. He looked at buying and selling shares in different companies and then examined the random behaviour of this portfolio as a whole.

His theory showed the importance of diversification in minimizing risk and maximizing returns. MPT introduced the concept of the efficient frontier, which helps investors construct optimal portfolios. Harry Markowitz won the 1990 Nobel Prize in Economics. We will be looking at MPT in a later chapter. There we'll see the pros and cons of this theory.

Efficient Market Hypothesis

There is a hypothesis in finance/economics that markets are efficient. Efficiency here means that all available information is reflected in asset prices. Obviously, this is called the Efficient Market Hypothesis (EMH). EMH was developed by Eugene Fama in the 1960s.

The hypothesis has important implications for investors, as it suggests that it is impossible to consistently beat the market through clever choices of stocks or market timing. Related to this hypothesis is the idea that share prices are predominantly random. And we are back at Bachelier.

> 'Too many people spend money they earned … to buy things they don't want … to impress people that they don't like.' Will Rogers

There are various forms of this hypothesis, depending on what information is included under 'all available.'

Whether the EMH is a good representation of real-world financial markets has been debated since it was first published. On the one hand there are countless people, and computers, looking at information that might help determine which shares to buy and sell that one can imagine that any mispricings won't last long. On the other hand, the vast majority of traders are not immune to behavioural effects such as herding, irrationality, overconfidence, etc. that will allow markets to be inefficient for quite a long time. One fun aspect of all this is that EMH says it's not worth analyzing the available information because it's already been done for you, it's already in market prices, but then if *no one* bothers to look at the information the market cannot be efficient, so you should look at information to gain an edge ... which makes the market efficient ... which means don't bother ... so the markets are inefficient ... And so on *ad infinitum*.

Personally, I'd say that financial markets are efficient, except for when they aren't.

Option Pricing Theory

We'll be seeing options in detail later. For now, you just need to know that they allow you to buy specified shares in the future but without being committed. As such one would expect that the value of these options would be closely linked to the value and future prospects of that share price. But can we theoretically model that relationship, between a share and its options?

For a long time no one knew how to value them. And a lot of clever people tried. Bachelier, yet again, was close but missed a crucial step. However, in 1973 Fischer Black, Myron Scholes and Robert C. Merton wrote two academic papers showing not only how to value them but also how to construct a simple portfolio of a share and an option that had zero risk. Scholes and Merton received the 1997 Nobel Prize in Economics, Fischer Black had died in 1995. Those are the same two Scholes and Merton who founded LTCM mentioned above.

The Black-Scholes model is probably the most accurate model in all of finance and economics, and thanks to it the market in options is in the quadrillions of dollars! On a historical note, Ed Thorp derived what is now known as the Black-Scholes option pricing formulae in the late 1960s, but never published them, instead using the ideas to make money! (Ed Thorp is an interesting character. He built, along with mathematician Claude Shannon, the world's first wearable computer, used to win at roulette. And he figured out how to win at the casino card game Blackjack by using the optimal strategy and card counting.)

> 'There is a very easy way to return from a casino with a small fortune: go there with a large one.' Jack Yelton

SOME JARGON

This chapter is quite long. I've put in it lots of descriptions of the world of finance. You might find yourself referring back to this chapter occasionally.

Interest: So now you know what money and banks are. Let's put some of that money in the bank. Why would you do that, why trust your hard earned to an anonymous institution? Several reasons. First, it's safer than keeping it at home, traditionally under the mattress. Even in the unlikely event that your bank is robbed, they have a record of how much money you gave to them and so you won't lose it. It's not as if every dollar bill in their keeping has your name on it. (I'm assuming your name isn't Washington.) Second, they give you more back than you gave them. This is called interest. You give them $100 and a year later you can take out, say, $104.30. That $4.30 is the interest. It's usually quoted as a percentage. So that's 4.3%. And quoted over a period, so 4.3% *per annum*, that's every year. Why do they give you back more than you gave them? A simple explanation is that they can use that money more profitably, for example lending it out so that they are the ones earning (even higher) interest. Different banks offer different interest rates. Lots more about interest coming later. Another reason for putting money in the bank is to build up a record with the bank of being a reliable, creditworthy, customer. One day you'll be asking that same bank to lend you money to buy your first home.

> '**Try to save something while your salary is small, it's impossible to save after you begin to earn more.**' Jack Benny

Fixed and floating rates: When you put money in the bank there is a choice to be made. Do you give the bank your money under the condition that you can take it out at any time? Or do you lock it away for a fixed period? If the bank is going to let you have the money back whenever you want it then the interest they give you will be floating. That means that the rate of interest can change at any time. Or you can specify a period. In that case the interest rate is set or fixed at the start, you know exactly how much you will be getting. Of course, if you suddenly need access to the money then you are stuck.

Base rates: If you follow the news then you might see reports of how central banks change an interest rate called a 'base rate' or 'bank rate' every month or so. This bank rate is the interest rate at which commercial banks borrow and lend their excess reserves to each other, usually overnight. What do I mean by reserves? The amount of money that banks have varies constantly as money is deposited and loans are made. But in order to maintain the stability of the banks they are typically legally required to have a certain amount in reserve, hence the frequent moving of money back and forth for short periods. The rate set by the central banks will then in turn affect the interest rate given by retail banks and the interest they charge on loans.

> 'Always borrow money from a pessimist; he doesn't expect to be paid back.' Anonymous

Maturity: Suppose you put money in the bank, locking it away for two years. You'll get paid interest, perhaps every month, perhaps once a year. But you won't be able to access that money for two years, until what is called maturity. Maturity is the period of a loan, whether it's you lending the bank money (your savings) or you borrowing from them to buy a car.

Principal: The amount that you lend the bank or they lend you is often called the principal. Calculations of how much interest you receive or pay are based upon the amount of the principal.

Mortgage: A mortgage is a loan from a bank that gives you enough to buy a home. In return for lending you the money the bank expects you to pay them more than they lent, this is interest again. Of course, if you don't pay them then they'll take your home away.

Bond: When you give your money to a bank for a fixed term what you've just bought is often called a bond. But bonds can also be issued by governments or corporations. Governments and corporations raise money for various projects by issuing bonds. As with the bank, a bond is usually for an agreed principal, agreed interest rate and agreed maturity. (The interest payments with bonds are also called coupons.) Each bank, corporation or government will offer different terms, such as the amount you can invest, the principal, and for how long. Importantly though, they will also offer different interest rates. And the rate of interest will depend on how creditworthy is the issuing institution, whether they are likely to default.

> 'Neither a borrower nor a lender be.'
> Polonius in *Hamlet* by William Shakespeare

Default: Default is failure to meet a legal financial obligation such as payment of interest or return of a loan. And there are usually consequences. Don't pay your monthly mortgage payment and you could lose your house. If a bank doesn't let you take out your money it could turn into a run on the bank. If a company doesn't pay a coupon on a bond it has issued this will affect how investors see the future of that company, expect a fall in its share price. If a country defaults on loans it will have a big impact on its currency, interest rates, and on its international standing. Likelihood of default will affect interest rates. Lend your money to the US government by buying one of their bonds and you'll get a certain interest rate, lend it to some bloke you've just met down the pool hall and you'd better get a much higher rate of interest. You might never see him again!

Credit risk and credit rating: Credit risk is associated with that risk of default. Often it is treated as random, meaning that one would refer to a certain percentage probability of default in the next year, say. This assumption of randomness is common in financial modelling, even though default is often a business decision. There are several companies such as Fitch and Moody's whose business it is to analyze the creditworthiness of businesses, and entire countries, and estimate probabilities of default. They will then give the company or country a rating of the type AAA, top notch, or BB+, speculative, etc.

> **'Inflation is when you pay fifteen dollars for the ten-dollar haircut you used to get for five dollars when you had hair.'** Sam Ewing

Share: Companies raise money for expansion, new ventures, etc. not just by issuing bonds but also by issuing stocks and shares. When you buy a bond from a company you are lending them money. But when you buy stock or a share in that company it is a part ownership of that company. Albeit usually an extremely tiny part of that company. Equity is the total share ownership in the company. You would buy shares in a company if you believe that the company is going to be doing well, and that the share price will rise giving you a financial profit. But you also make money from shares because they usually pay dividends. Shares are the subject of a later chapter.

Dividend: A dividend is a payment, often every three months, to shareholders in a company, paid out of their profits. It is to encourage people to invest in the company. It is also a way for the owners of the company to pay themselves, as opposed to, say, a salary. Dividends might be taxed differently from salary, but that depends on the tax jurisdiction.

Exchange: An exchange is a marketplace where financial instruments, such as shares, are traded. In London these grew out of coffeehouses but have now evolved to be fully electronic. If a company wants its shares to be traded it will need to be listed on an exchange.

Index: 'The Dow Jones index [fell, rose, plummeted, hit record highs/lows, ... insert yesterday's news here],' is a common lead story for many news outlets. The Dow Jones Industrial Average, its full name, is essentially the average of the share prices of 30 important US companies. (It's a tiny bit more complicated than that, this average gets divided by a certain number, the divisor that gets adjusted every time a company in the list of 30 gets replaced by another, it's just so that the index doesn't suddenly jump in value.) Indexes, or indices, are a way of gauging the overall behaviour of financial markets as a whole, or specific sectors of it. Common indexes are:

Index name	What it represents	Number of constituents
DJIA	Prominent companies listed on US exchanges	30
S&P500	Large companies listed on US exchanges	500
NASDAQ Composite	Mostly Information Technology companies listed on US exchanges	Approx. 2,500
FTSE100	Largest companies listed on London Stock Exchange	100
Nikkei 225	Companies listed on Tokyo Stock Exchange	225
FTSE All World	Global equities	Approx. 4,000
S&P GSCI	Commodities (see below)	Approx. 24
US Dollar Index	Value of US dollar in terms of other currencies	6
FTSE World Government Bond Index	Global fixed income (bonds)	20 countries

Markets: We've seen examples of two financial markets so far, the bond market, also called fixed income (because you know how much you will earn from an investment), and the equity market for shares. There are many other markets for many other kinds of financial instruments. For clarification, a financial market doesn't look anything like the market you visit on a Saturday morning to buy large bags of mixed nuts or handmade toe rings. The experience of a financial market these days is via a computer screen, or for the professional, multiple screens. Here are three more examples of financial markets.

Foreign exchange: Foreign exchange, forex or FX, markets are where large amounts of currencies are exchanged, as opposed to the high-street kiosk where you might get the relatively small amount of pesos for your summer holiday. Foreign currencies make up the largest financial market in the world. Currencies are exchanged in pairs, you might buy US dollars with pound sterling, or Japanese yen with Albanian lek. A country's currency can either be fixed or free floating. A currency that is fixed is tied to some other currency at a rate that doesn't change. The Saudi riyal is pegged at 3.75 to the US dollar. Free-floating currencies change value relative to other currencies in response to factors such as supply and demand, interest rates, etc.

> 'I'm only rich because I know when I'm wrong ... I basically have survived by recognizing my mistakes.' George Soros

Commodities: Commodities are raw materials used in production or for consumption. Examples are gold, oil, coffee, pork, wheat, corn, natural gas, copper, rubber, and so on. Commodity markets allow producers of the raw materials and their consumers to trade. Of course, much of the trading is also done by finance professionals trying to make money by correctly forecasting the direction of commodity prices, with no intention of actually taking delivery of a truckload of pork bellies. Trading happens via contracts known as forwards and futures. Forwards and futures are financial agreements to buy something in the future. The commodity, the underlying asset, isn't delivered until later. With forwards, money isn't exchanged until maturity of the contract either. Forwards and futures are contracts locking you into buying something later, at an agreed upon price. But you can close out the position, so you won't have to take delivery, by taking the opposite position, for example entering into a contract to sell, which cancels out the contract to buy.

Options: If forwards and futures lock you into buying or selling at a later date, options give you, well, the option to buy or sell a specified share at a later date. You can then buy/sell that specified share, called the underlying asset, at that date if you want. That later date is called the expiration date, it's the equivalent of the maturity with a bond.

The amount you would pay to buy or to sell the share is called the strike or exercise price. A contract to buy the specified share is called a call option, and to sell is called a put option. Options are discussed in detail later.

Long and short: I usually talk about buying shares for investment. But did you know that you can sell shares too? No, I don't mean sell shares that you own, that you bought last month. I mean selling shares that you don't own! Hmm. How can you sell something you don't have? Legally, I mean. Ok, how would you go about selling a lawnmower, say, that you don't own, and why would you do it? The answer to the second part is that you'd sell if you thought that the market price of lawnmowers was going to fall. Buying something is called going long, and selling something is called going short. You'd sell a lawnmower short, so you now have minus one lawnmower (bear with me here) and some cash from the sale, and then when the price of lawnmowers fell you'd buy one back, giving you a net position of zero lawnmowers (minus one plus one) and a profit being the difference between what you sold for and what you bought for. Neat! Unfortunately, it does look as though a spot of time travel is necessary, to sell something before you own it. How is this done in practice? Easy. All you need is a friendly neighbour from whom you can borrow a lawnmower in the first place. The neighbour doesn't need to know that you are going to be selling their lawnmower. As long as they get it (or exactly the same make and model) back later they will be blissfully ignorant. The same thing happens in the equity market, you borrow shares in order to sell them. Of course, this being finance there's no such thing as a free lunch (more on this later) so you have to make a payment in order to borrow the shares in the first place.

> **'I will tell you the secret to getting rich on Wall Street. You try to be greedy when others are fearful. And you try to be fearful when others are greedy.'** Warren Buffett

Return: Return is the growth in an investment usually expressed as a percentage. That's typically what you are looking for in any investment, a nice big return. If you've invested $500 in a share and it rises to $525 then you've made a profit of $25. That would be expressed as a 5% return, (525 - 500)/500. If you've invested in a bond with a fixed interest rate then you'll know exactly what return you will be getting. On the other hand, putting that $500 into a share will expose you to ...

> 'My formula for success is rise early, work late and strike oil.' JP Getty

Risk: Risk is the uncertainty associated with investments. Ok, this time the $500 became $525, but if you'd put the money into some other share then the $500 could now be just $400. And even the $525 that the share is worth now could disappear by tomorrow morning. Should you sell now to lock in that $25 profit or hold the share in the hope of further gains? How do you feel about taking risks? Decisions, decisions. We will be quantifying risk soon, measuring it mathematically using share price data.

Expectation: Key to much of the mathematics of finance are probabilistic ideas. In particular we need to know about averages, or expectations. What do we think is the expected return on a share? We know that owning the share is risky, but what is going to happen to that share on average? Some days it goes up in value, some days it goes down, where might we expect it to be in, say, one year? The expected return is sometimes called the expected growth. The mathematics of expectations is coming up soon.

Standard deviation: I've mentioned risk. Can we quantify it? Again, we have an idea from probability, the standard deviation. Standard deviation is a mathematical measure of how far numbers vary from their average. So we can talk about the standard deviation of returns. The mathematics is covered later.

Volatility: People in finance don't often talk in mathematical terms deliberately. But one thing they will always mention is volatility, and that is simply a rescaling of the standard deviation of returns. I think it's obvious that the longer you hold a share the greater the risk. The range over which the share price could move is going to be much greater over a five-year period than over five days. To simplify matters if you want to describe how much a share is bouncing around, its risk, then use a quantity known as volatility. This is just the standard deviation scaled to represent what the standard deviation would

> **'Annual income twenty pounds, annual expenditure nineteen and six, result happiness. Annual income twenty pounds, annual expenditure twenty pounds ought and six, result misery.' Wilkins Micawber in *David Copperfield* by Charles Dickens**

be over a one-year period and quoted as a percentage. It's a convention so that you don't need to specify a time frame. You might say, 'XYZ's volatility was 27% recently, but it's calmed down and is only 18% now.'

Correlation: Few people invest in just a single company. Most investors have a portfolio of shares, meaning that they invest in several different companies. The reason behind this is captured in that saying, 'Don't put all your eggs in one basket.' By holding shares in several companies you are reducing exposure to any single stock. If one share price falls it's not the end of the world. This is diversification in action. Calculating the expected return on a portfolio of shares is mathematically simple, just add up the expected returns on the individual shares weighted according to how much you hold in each, details later. But measuring the risk or standard deviation of a portfolio is harder. And that gets us onto the subject of correlation. Roughly speaking, with the mathematics later, when two stocks are perfectly correlated it means they move in lock step, they go up or down together. If perfectly negatively correlated then when one moves up the other down, and vice versa. Zero correlation means that there is no simple linear relationship between them. If your portfolio consists of assets that are perfectly positively correlated then you aren't really diversified at all! (I'm speaking a bit colloquially here when I say things like 'one moves up, the other down.' Technically you measure correlation after you've taken off the expected move, and scaled with the standard deviation. Don't worry precisely what this means yet, I just wanted to cover myself in case of complaints about me writing rubbish!)

Risk averse, risk neutral, risk seeking: It's perfectly normal to be wary of financial risk. No rational person wants to lose money unnecessarily, do they? Usually, people take financial risk because they expect it to pay off, that the return justifies taking the risk. Someone offers to bet $1 on the toss on a coin, you get $1 if the coin lands heads, you pay out $1 if tails. Would you take that bet? You might for a bit of fun. Make that $10,000 with the same odds and not many people would. That illustrates the idea that the value of financial risk is important in making investment decisions. What if I say that you'll get $10,000 if a die is rolled and lands on 2–6, but if it lands with 1 on top you pay $10,000? Now you might take the bet if you are rich enough, or get some friends together to come up with the money in a syndicate. And that illustrates the idea that the odds matter, that what you expect to happen might outweigh the risk involved. Generally speaking, people are risk averse, meaning that they need to be compensated for taking risk, that the expectation must be sufficiently positive to outweigh the standard deviation. But there are times when people take risks with a negative expected return. One obvious example is the lottery ticket. You pay $1 for a ticket but the statistics say that you expect to get back 80 cents perhaps. Now lottery tickets don't pay 80 cents, if you win you'll get either $10, or $100, or … $100,000,000! But the majority of ticket holders won't see a cent. The mean of all of those payoffs, multiplied by how many of each there are, is $0.80. Making a bet that has a negative expected return is considered to be irrational, it is risk *seeking*. However, if you are poor but need an expensive life-saving operation and don't have access to free healthcare then it might be very rational to buy that lottery ticket. Between risk averse and risk seeking there is a category of investors who are risk neutral. This simply means that they would make investment decisions based solely on expectations without any concern for risk. The concept of risk neutrality will be important when we get on to valuing options.

> 'Rich people have small TVs and big libraries, and poor people have small libraries and big TVs.' Zig Ziglar

Arbitrage: In its purest form arbitrage is the simultaneous buying and selling of the same asset on two different exchanges to make a profit. That profit comes about because you've spotted that exactly the same asset has one price on one exchange and a different price on another, so you buy the cheap and sell the expensive, thus making that profit. (You have to allow for bid and offer prices, bid being the price that a buyer is willing to pay for something and offer or ask being what they are willing to sell for.) Other arbitrages might exist because you find that two different portfolios give you the same payoff at some date in the future, yet the two portfolios cost different amounts to set up today. So buy the cheap one and sell the expensive. At that future date the two payoffs will cancel, but you will have made money on the deals upfront. Arbitrage opportunities are not common because the minute one arises there will be some experienced trader (or computer!) who takes advantage of it, causing prices to move via supply and demand so that the arbitrage disappears. Indeed, the absence of arbitrage opportunities plays an important role in valuing options. 'There's no such thing as a free lunch.' Another nice example of arbitrage comes from the foreign exchange markets. Suppose you see the following exchange rates: EUR/USD = 0.9497, GBP/EUR = 0.8789, and USD/GBP = 1.2398. Can you profit from these rates? This is how it goes. Take a million US dollars and buy euros, that's €949,700.00. Use that €949,700.00 to buy 0.8789 x 949,700.00 pounds sterling, that's £834,691.33. Now take £834,691.33 to buy 1.2398 x 834,691.33 US dollars, that's $1,034,850.31. Well done, you've just made $34,850.31 out of thin air! That's called triangular arbitrage. In practice the spread of prices, bid-offer spread, means that such arbitrages rarely exist in practice these days. Any arbitrages are likely to involve financial instruments that are more complicated than shares or exchange rates. Sometimes the arbitrage is probabilistic, meaning that you'll *probably* make a profit but it's not guaranteed, that's statistical arbitrage.

> 'There's no such thing as a free lunch.' Milton Friedman

Hedging: Hedging is a very important strategy that is meant to reduce risk by exploiting correlations between assets. Suppose you hold one investment that you deem to be too risky, you could buy a second asset that has a negative correlation with the first, so reducing the overall risk. Or sell a second asset with a positive correlation. Of course, the downside of reducing the risk is that you might also be reducing the expected return. We will see a very nice, and Nobel Prize winning, hedging strategy when we come to look at options.

Leverage: Leverage is when you borrow money to invest. For example, you've got $10,000 in savings, and want to invest in a stock that you are sure is going to double in value over the next six months. If you are right, then you will make a profit of $10,000. But if you are really, I mean really, sure about this doubling then you could borrow another $10,000, heck, make it $100,000. Now invest the whole $110,000. After doubling, you will have $220,000. Pay back the $100k, and whatever interest you had to pay on the loan, and you will have made almost $120k. Wow! However, if the stock doesn't live up to expectations but halves instead, then the $110k will becomes $55k, and you will end up owing money you don't have, $45,000 plus interest. Leverage is risky.

Hedge funds: Hedge funds manage money for private clients, investing in a wide range of financial products. But these clients are not your average man in the street, these clients are the extremely rich, high net worth individuals. There's usually a minimum investment requirement. Hedge funds can invest in all sorts of financial instruments, including the quite exotic. They earn money by charging management and performance fees. Clients might pay, say, 2% of their invested money *per annum*, and then 20% of the profit that the hedge fund makes for them. Even though they are called hedge funds they don't necessarily do all that much hedging. Instead, they will often employ leverage. That's one of the reasons for the collapse of LTCM, mentioned earlier. However, some have been so successful that they no longer take clients' money but only manage the money of their owners and employees. Hedge funds can be very secretive about what they do. Sometimes their websites have no information at all, only a discrete login for their clients.

> 'What is a cynic?' ... 'A man who knows the price of everything and the value of nothing.'
> Cecil Graham and Lord Darlington in *Lady Windermere's Fan*,
> Oscar Wilde

> **'Dogs have no money. Isn't that amazing? They're broke their entire lives. But they get through. You know why dogs have no money? No Pockets.'** Jerry Seinfeld

Fundamental analysis: You want to invest in the financial markets to get a nice return on your savings. But we know that many financial instruments are risky. Take equities for example, they could go up nicely or they could plummet. How do you know which shares to invest in? You could throw a dart at a copy of the *Financial Times*. Or you could do some analysis. An obvious analysis to undertake is to examine in detail the company you fancy. This is called fundamental analysis. If the company has shops then you visit those shops, are the customers satisfied? Are there customers?! Who is on the board of the company? Are they there because of their experience and skills, or because they went to the right school? Does the company have valuable patents, but are they expiring soon? Open up the annual accounts, examine the balance sheet. Oof. All that is a lot of hard work, it could take weeks. But it must pay off, right? Surprisingly the answer is no! When you buy shares in a company you are hoping that the price rises so you can sell and pocket the difference. But the price at which you can later sell depends on market sentiment. And that market sentiment might not be totally rational, other participants might not have gone to the same analytical lengths as you. The problem here is best conveyed by some very insightful quotations from the economist John Maynard Keynes (pronounced like canes), 'The markets are moved by animal spirits, and not by reason.' And most famously, 'Markets can remain irrational longer than you can remain solvent.' If you take nothing else from this book, take away those insights.

Technical analysis: If fundamental analysis is of less use than you might expect, what else can be done? And something less time-consuming, please.

Well, there's technical analysis. This is the opposite of fundamental analysis. Where fundamental analysts spend ages going through a company's accounts, the technical analyst or chartist just draws a few lines on a graph. Where fundamental analysts want to know everything about a company, the technical analysts claim that everything is contained within the plot of share price versus time, they don't even want to know the name of the company! The chartists look for certain shapes in the graphs, saucer bottoms, head and shoulders, triple tops, and they plot various curves, moving averages, Fibonacci series, Bollinger bands, and so on. The scientific evidence for the efficacy of technical analysis is mixed, but tends towards the negative. However, this is the chartists' own fault. If chartists could all just agree on a *single* indicator then it would become self-fulfilling! All chartists would, as one, say, 'Buy stock XYZ!' And people would buy and the share price of XYZ would rise. As predicted! That's my top tip for chartists!

Quantitative analysis: If the above two methods of analysis are disappointing then we still have one more methodology up our sleeves, quantitative analysis. This follows on from the idea of the Efficient Market Hypothesis. Trying to predict anything might be hopeless according to this hypothesis, but that doesn't mean we can't have a mathematical model. It just means that if you are trying to decide which shares to invest in then you need to have some probabilistic model for share price behaviour, a model for its randomness. And that amounts to having a model a of share's expected growth and volatility.

> 'There were times my pants were so thin I could sit on a dime and tell if it was heads or tails.'
> Spencer Tracy

CAREERS IN FINANCE

I'm going to describe a selection of finance jobs, and I'm going to distinguish between technical jobs and non-technical jobs. This division is based on how much mathematics or computer science knowledge is required. Having to use spreadsheets does not count as being sufficiently technical here!

Non-technical Careers

Accountant: An accountant tracks and records the financial goings on of a company, partnership, sole trader, or individual. Incomings, outgoings, what expenses are allowed, depreciation of assets, reconciling bank statements, how much tax is to be paid, and so on. The mathematics is probably not difficult – although I struggle since it's all numbers and no x's and y's! – but the job requires a talent for keeping track of details and keeping up to date with changes in taxation, and other regulations. To be an accountant you usually need certain professional qualifications.

> 'The trick is to stop thinking of it as 'your' money.' IRS auditor

Auditor: An auditor is similar to an accountant but instead of compiling data they review the accounts of companies and other organisations making sure that they are correct.

Financial Analyst: A financial analyst examines financial documents to forecast companies' future prospects. These are the people who do the fundamental analysis mentioned earlier. They will have to make presentations based on their analyses. One subtle point, even though fundamental analysis might not be as helpful as you'd hope when it comes to forecasting share prices, it is still very important for the running of companies, in making important decisions.

Financial Advisor: A financial advisor gives advice to clients about how to invest their money, whether to put it into simple products, or more complex and risky, and in what proportions. They will usually conduct some assessment to figure out your risk profile. Are you risk averse or a risk taker, do you need easy access to your money or are you happy to lock it up for a while?

Stockbroker: Stockbrokers buy and sell bonds, shares and other products on behalf of their clients. They will also offer advice on these products. They are middlepersons, if you like. Stockbrokers earn money through commission and advisory fees. As such they don't really have 'skin in the game,' if they give poor advice then their downside is only in reputation and job security. This is a job for a people person.

Securities Trader: Securities traders also buy and sell financial products but they usually do so on behalf of the company they work for, which may be a bank. Although some people might trade privately for themselves. Traders typically get compensated in a large part via bonuses, rather than a base salary. And that bonus can get very large indeed if the trader is very successful. These traders might use mathematical models to help them make trading decisions, but they aren't usually required to understand the details of these models. This can be a very stressful career choice, with little job security.

> **'What's worth doing is worth doing for money.'**
> **Gordon Gekko (Michael Douglas) in *Wall Street***

Venture Capitalist: Venture capitalists look out for companies that need investment money for growth. They will invest in the company in return for some ownership and then provide advice and often board members. Venture capital, VC, companies might use their own money or money that they have raised from their own investors. Their profit comes when the company that they have invested in gets sold. A career in venture capital can be lucrative.

Investment Banker: An investment banker raises capital for their clients by selling equity in the company and issuing debt, selling bonds. They also help in mergers and acquisitions. They will have to analyze economics and markets, and advise on risks. They will work with financial models, but again won't need to understand the nitty gritty. They will be paid in part via a base salary but also with a, potentially much larger, performance bonus.

Chief Financial Officer: The Chief Financial Officer (CFO) of a company is the top dog in a company when it comes to matters of finance, and thus one of the most important people in any organization. They will analyze risks, make financial plans, oversee budgets, analyze data, keep records and make reports. They will be paid in salary, company shares, and performance bonus.

Technical Careers

Actuary: Actuaries analyze financial risk within businesses. They use statistics, probability and other mathematics. Insurance companies in particular employ actuaries to study matters of life and death, or rather the probabilities thereof. But they can be found anywhere that requires an analytical approach to risk. Government actuaries advise the government and public sector departments and businesses.

> '**I made my money the old-fashioned way. I was very nice to a wealthy relative right before he died.**' Malcolm Forbes

Data Scientist: Data scientists collect and analyze historical data, and build predictive models based on that data. The process of collecting data involves data mining which is the collection of the relevant data in the first place, then there's data cleaning which is fixing incorrect data, filling in missing data, removing duplicates, etc. They will use statistical methods, machine learning tools, and so on. This is a field that is very rapidly changing. More and more data are being collected and stored, thanks to increasingly computer capabilities and the internet. And the rapid development of artificial intelligence means that data scientists have to do a lot of running to stay on the same spot.

Portfolio Manager: Portfolio managers decide how to allocate money across different assets, whether for their firm or for clients. They might have to analyze the equity market, bond market, commodities, property, etc. And in deciding how to allocate money across these different markets they will have to look at both the details of individual assets as well as the big economic picture. An important part of this is understanding diversification, and this means understanding theoretical financial concepts. They can work for investment banks, or in venture capital or for hedge funds.

Risk Manager: Risk managers analyze risk and manage it. Obvs. This can get very technical within banks because they have to understand many financial products, how they are related to each other, what hedging there is. They will have to measure value at risk, which is the amount that could be lost with a certain probability over a certain timeframe, and also stress test portfolios to see what might happen in the event of extreme market moves and crashes. They need to understand and be able to model not just the future behaviour of assets but also the statistical properties of the assets such as how volatilities might change.

Financial Software/Quantitative Developer: And now onto a job involving computer programming. The complexities of financial products, bonds, equities, options, etc. require software that can analyze their values, their risks, their correlations, future scenarios, take in data from third parties, give out data, not to mention be used by many people with different levels of sophistication throughout a bank. The software needs to be accurate, fast, user friendly, and constantly updated and added to in response to new products and new financial regulations. It needs to fit into the banks' other systems, such as spreadsheets, and work flawlessly with various operating systems. Developers might work within a bank. But there are also many private companies that provide such software for the banks. Wherever you work you need to be up to date with the latest financial theories and products. If you like coding this could be the job for you.

> '**Money is not the most important thing in the world. Love is. Fortunately, I love money.**' Jackie Mason

Quantitative Analyst/Quant: Finally, the Quant. This is the most mathematical of all the jobs described so far. Quants are in great demand when it comes to analyzing the more complex financial instruments, such as options. They don't just need to know all about the latest financial theories but if they are lucky to work in the right environment they will get to develop those theories themselves. Sometimes these theories and mathematical models remain secret, but sometimes the quants get to write academic papers and speak at conferences about their new models. As well as understanding the niceties of statistics, probability, calculus, etc. they also need to be good computer programmers. In banks they might be valuing new products, and figuring out how risky they are and how to reduce that risk. In hedge funds they might be looking for arbitrage opportunities and mispriced assets.

Here's my financial advice: Do not google the salaries of most of the above, you will get very envious. (Ok, you can safely google accountant.)

'Never confuse the size of your paycheck with the size of your talent.' Marlon Brando

SOME MATHEMATICS

This chapter is about some basic concepts in probability. We'll be needing these ideas when we get on to representing share prices, portfolios, etc. mathematically. As I mentioned, we can approach the analysis of share prices in many ways, such as fundamental analysis and technical analysis. But the methods that have been most successful for the last few decades also happen to have the most mathematics and treat share prices as random.

Expectation

A starting point for anything involving randomness is to look at what happens on average. There are several mathematical ways to measure averages of numbers. There's the mode, median and mean. The mode is the most common number in a list, the median is the number in the middle when they are put in order, and the mean is the average calculated by adding up all the values and dividing by how many numbers there are. But you knew that. If you have the numbers 1, 4, 9, 2, 3, 2, 5 then the mode is 2, the median is 3 and the mean is 26/7. The mean is also called the expectation or expected value, and is the average that we'll be working with from now on. Although it's harder to measure than the mode or median it has nice mathematical properties.

Let's introduce some symbols. Write the list of numbers as $x_1, x_2, \ldots, x_n$. So the expectation is just

$$\text{Expectation} = \bar{x} = \frac{1}{n}(x_1 + x_2 + \ldots + x_n) = \frac{1}{n}\sum_{i=1}^{n} x_i.$$

That was easy. That's where I got the 26/7 from. I've also introduced the symbol $\bar{x}$ for the expectation, I will use that occasionally below.

But that was just a list of numbers. So far, I haven't mentioned probability. And that's rather important when dealing with things that are random. I've just given you the list and a calculation to perform to get the average. Probabilities would come in if those numbers were drawn from a hat full of random numbers, or we could talk about rolling a die …

If I rolled a die many times and made a list of all the numbers that I rolled then I could do the same expectation calculation as above. That's quite a lot of effort, rolling enough times to get a sufficient list of numbers to give an accurate measure of the average. However, for an unbiased die we can get the same result by an equivalent calculation without doing any rolling at all:

$$\frac{1}{6}(1 + 2 + 3 + 4 + 5 + 6) = 3.5.$$

That's the expected value. See how I've used information about the probabilities of each number in the calculation? That's 1/6 for each number.

Things get more interesting if our die is biased, so that the probabilities are not all the same. Then *if we knew* the probabilities of getting each number, we'd do a similar calculation to find the expectation but using different probabilities. Generally, we'd have

$$\text{Expectation} = \bar{x} = \sum_{i=1}^{n} p_i x_i = E[x].$$

I've multiplied each number by its probability, that's the p_i, and summed. There's no $1/n$ in front because that is in the p_is. And because the p_is are probabilities of discrete events they cannot be negative and must sum to 1.

$$\sum_{i=1}^{n} p_i = 1.$$

(And I've introduced another way of writing expectation, that's the $E[x]$.)

In practice, if we had a die that we thought was biased, maybe it had a bit of lead fishing weight underneath the one, opposite the six (we've all done this, haven't we?), then we would be unlikely to know the exact probabilities, the p_is. We'd be back at rolling the die many times to estimate the average.

Properties of expectations

You have one random number, x, and another random number, y. The first number has an expected value, $E[x]$, and the second, $E[y]$. If you add these two random numbers, $x + y$, I wonder what the expectation of that sum would be. It could not be simpler, it's just $E[x] + E[y]$. We write

$$E[x + y] = E[x] + E[y].$$

In words, the expectation of the sum of random variables is the sum of the expectations of each random variable.

And this works for adding up any number of random numbers. They don't have to be independent of each other. And they can be from different distributions. One could be the result of rolling a die, and the other from spinning a roulette wheel. Simple! (This is the nice mathematical property of expectations, means, that I mentioned earlier. Unfortunately, there's no nice formula like this for modes and medians.)

For example, toss two fair coins. For each coin, heads counts as plus one, tails is minus one. The expected value for both coins is zero. Add the two together and the expected value of the two coins is zero plus zero, equals zero.

Standard Deviation

The expectation is a measure of what might be, ahem, expected to happen, but is there any way to measure how much variation there is about this average? Yes, there are a few. One measure is the range. With the above list of numbers, the range is 8, from 1 to 9. Should we use this to measure how share prices might vary? Well, it might be a bit too extreme. For example, a share price could fall all the way to zero! But the range doesn't measure how likely that is. Another measure is the standard deviation. To define and calculate this we first need a quantity called the variance.

Take all the differences between the numbers in the list and the expectation, square them, and average:

$$\text{Variance} = \frac{1}{n}\sum_{i=1}^{n}(x_i - \bar{x})^2.$$

For our list of numbers this is

$$\tfrac{1}{7}\left(\left(1-\tfrac{24}{7}\right)^2 + \left(4-\tfrac{24}{7}\right)^2 + \left(9-\tfrac{24}{7}\right)^2 + \left(2-\tfrac{24}{7}\right)^2 + \left(3-\tfrac{24}{7}\right)^2 + \left(2-\tfrac{24}{7}\right)^2 + \left(5-\tfrac{24}{7}\right)^2\right) = \tfrac{304}{49}.$$

But this is not quite the measure we want. There's a problem with this variance as a measure of variation about the mean. Suppose that the numbers in our list represent prices measured in dollars, $. This variance would then have units of dollars squared, ^{2}. However, any measure of variation ought to have the same units of measurement as the thing we are measuring, otherwise we'd be comparing apples and Ford Escorts. To get back to $, to get a number that measures deviation about the mean, we need to take the square root. And that's the standard deviation. For our numbers that would be $\sqrt{304/49} = 2.491$. In symbols

$$\text{Standard deviation} = \sqrt{\frac{1}{n}\sum_{i=1}^{n}(x_i - \bar{x})^2}.$$

If we are dealing with random numbers for which we know the probabilities, such as with the fair die above, then the variance is

$$\text{Variance} = \sum_{i=1}^{n} p_i(x_i - \bar{x})^2 = \text{Var}[x].$$

This can also be written as an expectation:

$$\text{Var}[x] = E\left[(x - \bar{x})^2\right].$$

And then the standard deviation is still

$$\text{Standard deviation} = \sqrt{\text{Var}[x]} = \text{StdDev}[x].$$

Properties of standard deviation

The additive property of expectations is not shared by standard deviations unfortunately. However, there is something similar for variances but there's a technical requirement to make it work.

No, we can't just add up standard deviations. But we can add variances, in some cases. The technical requirement for being able to add variances of two random numbers is that *the correlation between them must be zero*. If that's the case, and we'll see correlation next, then

$$\text{Var}[x + y] = \text{Var}[x] + \text{Var}[y].$$

Toss two unbiased coins again. Heads means one, tails mean minus one. The mean for each coin is zero, so the variance for each coin is

$$\frac{1}{2}\left(1^2 + (-1)^2\right) = 1.$$

And the standard deviation, the square root of this, is one. But for two coins we can't add the standard deviations. The standard deviation for these two *independent* coins is *not* two. However, we can add the variances, giving a variance for the two coins as two, and so the standard deviation for the sum of the two coins is $\sqrt{2}$. For n coins the standard deviation would be $\sqrt{n}$.

Correlation

Here are two sets of numbers. Both range from 1 to 9. But they have different statistical properties. In particular Set X has a smaller mean but a larger standard deviation than Set Y. In a later chapter we'll be talking about how to measure means and standard deviations of share prices returns, and how larger expectations and smaller standard deviations are better.

	X	Y
	1	4
	4	5
	9	9
	2	1
	3	4
	2	5
	5	3
Mean	3.714	4.429
Std Dev.	2.491	2.259

When you have two sets of numbers, like the X and Y, you might wonder if there's any relationship between the two sets. Suppose they come in pairs, (1, 4), (4, 5), (9, 9), etc. For example, the first list is Person X's score on several Latin grammar tests and the second is Person Y's *on the same tests*. Person X has a mean score of 3.714, Person Y has a slightly larger 4.429, and they have similar standard deviations. But if Person X does well on one test does Person

Y do well on the same test? Correlation can be used to measure this. I'll go straight to the formula and then explain what it means. I'll use x_i to represent Person X's scores and y_i Person Y's.

First, we need the covariance:

$$Cov[x, y] = \frac{1}{n}\left(\sum_{i=1}^{n}(x_i - \bar{x})(y_i - \bar{y})\right).$$

Another way of writing this covariance is as an expectation:

$$Cov[x, y] = E[(x - \bar{x})(y - \bar{y})].$$

For our A and B data this covariance comes out at 1.796.

And then the correlation is

$$Correl[x, y] = \frac{Cov[x, y]}{StdDev[x]\ StdDev[y]}.$$

This correlation is a measure of any linear relationship between two random variables. Correlation can be from -1 to +1. It works out at 0.7327 for our numbers.

A correlation of 0.7327 is not a perfect correlation, that would be a 1. But nor does it show there's no linear relationship, that would be a zero. So there is some positive relationship there. However, you will often hear, 'Correlation is not causation.' This means that from this one number we can't tell whether X copies test results from Y, or vice versa, or whether their similar results are because of some common factor, such as some tests being easier than others.

> **Aside: A note on notation**
>
> I'm writing the expectation of a random variable x as $E[x]$. Sometimes you'll see $E(x)$, $Exp(x)$, $\mathbb{E}[x]$ or simply Ex. And similarly for variance, V, $\mathbb{V}$, or Var, and with $(\ )$ or $[\ ]$. It really doesn't matter. Personally I prefer to use the symbols that convey the minimum necessary, and no more. And I don't really like $(\ )$ since I use that for functions, whereas expectation and variance are technically functionals, which are a bit more complicated. But nothing you need to worry about now!

> **Aside: Don't be like UK politicians!**
>
> The Royal Statistical Society did a survey of UK Members of Parliament, asking them if a coin is tossed twice what is the probability of getting two heads. You know the answer, I am sure. It is 25%. The probability of the first head is 0.5, the probability of the second is 0.5. The tosses are independent so multiply. Simple.
>
> But MPs are not as smart as you. Only 40% of MPs got the answer right.
>
> Some MPs said the answer was one in three. They probably thought that a head then a tail was the same as a tail then a head. (If you thought this then think of tossing two coins, one silver and one copper.)
>
> A large numbers of the MPs gave the answer 50%. I've no idea what they were thinking!
>
> MPs of one main party were twice as likely to be correct as MPs of the other. I won't tell you which party is smarter!

Ok, that's enough of the preparation, now let's look at some mathematical finance in depth.

IN DEPTH: FIXED INCOME

Fixed income generally refers to financial instruments having known future payments, with known interest payments or coupons. They differ from those instruments whose future is uncertain, such as anything involving shares, commodities or exchange rates, for example. It's not quite that simple because interest rates themselves do change, and putting your money in an instant-access bank account does not give a guaranteed return.

You put $100 into a fixed-term bank account maturing in one year. A year later they give you the $100 principal back and an extra amount, the interest. If we ignore credit risk, that's the risk of the bank not paying you back, then these instruments are risk free and simple to understand. You know exactly where you stand, you know the return you will make. There are downsides though. One is that you can't access that money before maturity. Another is that interest rates might rise during the year and you will have missed out on a higher rate of interest.

Similar but different, is putting your money in the bank with no maturity, so that it is immediately accessible. You will still get interest, but it will be paid perhaps every month, and the interest rate will be different from if you'd locked up your money at a fixed rate. And that rate can always change, depending on such things as the rates set by central banks and supply and demand.

Locking your money away you get a fixed rate of interest. Instant access and you get a floating rate.

I've introduced you to the ideas of fixed and floating rates, now let's start on the mathematics. For this chapter my examples will be from the world of retail banking, just putting your money into a main/high street bank account, and getting fixed interest payments. But bonds issued by companies and governments have similar mathematics behind them. In the Appendix there's an introduction to the mathematics of floating rates.

The Mathematics

Basic calculations

We'll start with some numbers and then gradually move over to symbols.

Put $100 into a fixed-rate account with a maturity of one year and an interest rate of 3.8%. How much money do you get back after a year? Easy:

$$100 \times (1 + 0.038) = 103.80.$$

Turn that around. You put $100 in the bank and get $104.10 back after one year. What was the interest rate? Pretty obviously it was 4.1%.

Slightly harder, $250 in the bank now, at 5.5% interest. After one year that's

$$250 \times (1 + 0.055) = 263.75.$$

Harder still, but still pretty basic, $400 in the bank now, and you get $426 back after a year. What was the rate of interest? Call it x.

$$400 \times (1 + x) = 426.$$

Rearrange this to get

$$x = \frac{426}{400} - 1 = 0.065.$$

That's 6.5%.

Now a bit harder again. $300 locked away in the bank at an interest rate of 4.7% for six months. How much will you get back?

Aha, it depends!

It depends on what we mean by '4.7% for *six months*.'

Here we come to the concept of conventions used in finance when quoting interest rates.

The convention is to quote in *annualized* terms. This means we translate interest rates to *what they would be over one year*. So when I say 4.7% that would mean 4.7% over one year. *Even though the maturity is in only six months*. That would therefore become 4.7/2 = 2.35% over the six months.

The answer to the question is thus

$$300 \times (1 + 0.0235) = 307.05.$$

And now turn that around, $150 in a three-month bond returns $151.45, what was the interest rate? Use x to mean the *annualized* rate of interest. We have

$$150 \times \left(1 + \frac{x}{4}\right) = 151.45.$$

The quarter is there because the interest is paid after three months, a quarter of a year. Rearranged this becomes

$$x = 4 \times \left(\frac{151.45}{150} - 1\right) = 0.0387.$$

That's 3.87%, in annualized terms.

Simple and compound interest

My examples have had different amounts of principal, different interest rates and different maturities, but all had one thing in common, there was only one payment in each case.

Things get more complicated, and more interesting, when there is more than one interest payment during the life of the account. And this brings us onto the subject of simple versus compound interest.

Simple interest is when the calculation of the amount of each payment is based on the initial principal only.

Compound interest is when the calculation of the amount of each payment is based on the initial principal *together with all previous interest payments*.

Here is simple interest in action, $100 principal, simple interest of 10%, paid every year for 10 years. That's ten interest payments in all. Each payment is 10% of $100, that's $10, and there are ten of them, so your $100 becomes $200 at maturity, $100 in the principal and another $100 in interest. That's pretty good, no?

And here is compound interest using exactly the same numbers. After one year the interest on the $100 is $10. But the next interest payment is now based on the principal plus interest so far, that's interest on $110, not on the $100 only. That's interest of $11. So after two years the bond will have accumulated $21 in interest, which together with the initial $100, gives $121. Then after three years that's 1.1^3 x 100, and after ten years

$$1.1^{10} \times 100 = 259.37.$$

Which is even better than what you get with simple interest. That's interest on the principal, together with interest on the interest, and interest on the interest on the interest, ...

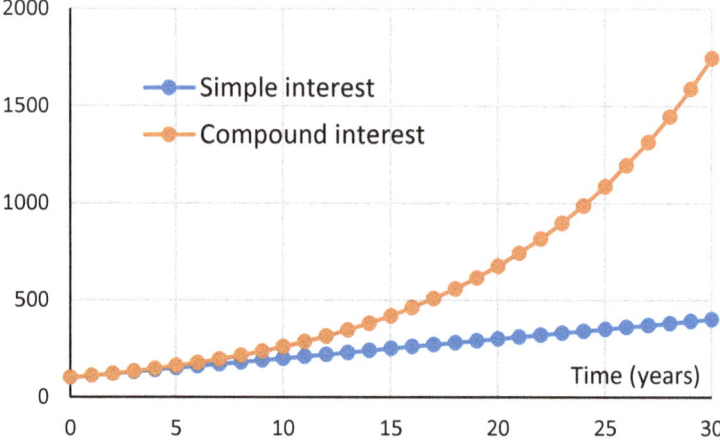

The plot shows how money grows with time with the two different types of interest. For short maturities the difference is small, but over longer periods compound interest far exceeds simple.

When you invest your money at a fixed rate in a bank they will ask you where you want your interest paid. They will ask whether it is to be paid into the same account as the principal, or into a different bank account. This doesn't matter if there is only one interest payment. But it does matter if there are several interest payments before maturity. If you choose to have it paid into the account where the principal is then it will probably be compounded. But you won't be able to access *anything* until maturity. If you choose to have it paid into a different account, perhaps into your checking/current account, then you will be able to access those interest payments, but future interest payments will be simple interest, only paid on the principal.

We will only be working with compound interest going forward.

> **Aside: The power of compounding**
>
> There are various stories or legends about the game of chess and its inventor, or perhaps just a good player, asking to be rewarded by receiving an amount of rice: 'Give me one grain of rice on the first square of the chessboard, two grains on the next square, four on the next, eight on the next, and so on, doubling the quantity for each subsequent square.' Well, $2^{64} - 1$, for that is the total number of grains of rice, is 18,446,744,073,709,551,615! And that's many thousands of times the world annual production of rice!

The mathematics of compound interest

Let's introduce some symbols to illustrate compound interest. Use the symbol P to represent the principal and r to represent the interest rate, in annualized terms. And let's have a maturity of time T, measured in years, and n interest payments every year. The interest payments are at an equal interval of $1/n$ years apart.

Remembering that the interest rate r is annualized we have to convert that, using the aforementioned convention, to the rate over one interval. That is simply r/n. (If this is not clear go back to the example where the interest payments were every three months, one quarter of a year.)

> **Aside: Annual Percentage Rate (APR)**
>
> The Annual Percentage Rate (APR) is the yearly interest generated by a principal that's charged to borrowers or paid to investors. It is expressed as a percentage that represents the yearly cost or income of the funds. It does *not* take compounding into account. The APR provides investors/borrowers with a single number that can be used to compare among different lenders, credit cards, or investment products.

After one interest payment the principal plus interest is equal to

$$\left(1 + \frac{r}{n}\right) P.$$

After two interest payments we have a total of

$$\left(1 + \frac{r}{n}\right)^2 P.$$

And finally at maturity, that's after nT interest payments, we get back

$$\left(1 + \frac{r}{n}\right)^{nT} P. \qquad (1)$$

Here's a plot showing what this looks like with more and more interest payments. Here r is 10%. When you look at this plot don't forget that those interest payments are getting more and more frequent, but the interest rate each time is getting smaller and smaller, r/n.

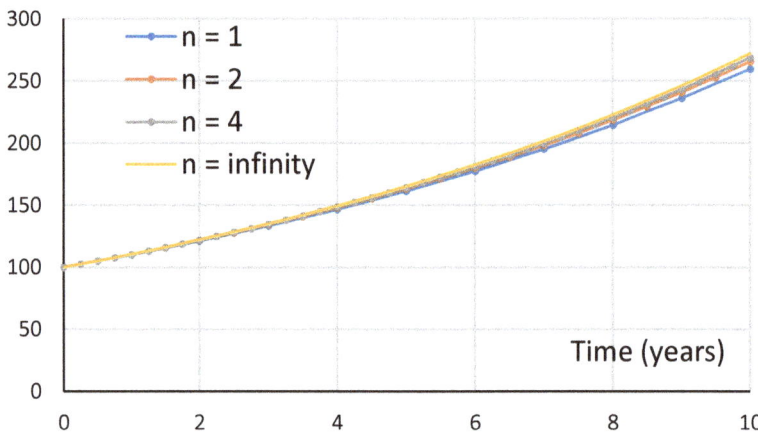

There are four lines in this plot, but they are very squished up. That's because the amounts of the interest payments are very similar. But they aren't exactly the same. Here is a close up of the top right of the plot, now you can see that the principal plus interest after ten years is slightly different depending on how many times you get the interest.

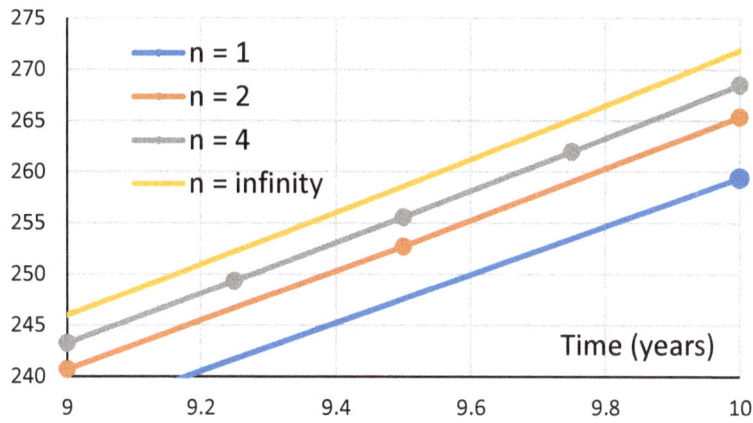

This is because one payment at a rate r once a year is not the same as two payments of r/2, is not the same as four payments of r/4, and so on. Again, this is the effect of compounding. The mathematics of n being infinity is discussed in the Appendix.

And this matters. Ok, maybe the $3 difference after ten years between interest paid twice a year as opposed to four times a year is not important, but traders who deal in these products are not working with principals of $100. No, they are dealing in many millions of dollars each trade. And then those differences become significant, they pay their bonuses!

> **Aside: A subtle point about conventions**
>
> Be careful when referring to interest that is not paid annually. Take the above formula and ask what is the total interest paid after one year when there are two compounded interest payments of r/2 each? The answer is
>
> $$\left(1+\frac{r}{2}\right)^2 - 1 = r + \frac{r^2}{4}.$$
>
> And this is not the same as an interest payment of r once a year! It's a little bit more. This applies to any interest payments that are paid more than once a year.
>
> Confused? Much of fixed income mathematics is about subtleties like this. Don't let it put you off though, just be careful about conventions.

Discounting

In my examples I've talked about investing, say, $100 and looked at how much that $100 becomes in the future after all the interest payments. What if we turn that around and ask how much must we invest now to get $100 at some time in the future?

This introduces the idea of the present value of that $100, and it is given by

$$\text{Present value} = \frac{100}{\left(1 + \frac{r}{n}\right)^{nT}}.$$

The multiplicative factor in this is called the discount factor:

$$\text{Discount factor} = \frac{1}{\left(1 + \frac{r}{n}\right)^{nT}}.$$

It nicely describes the concept of the time value of money, that money in the future is worth less than money in your hand!

Backing out an interest rate

We've backed out an interest rate when there is one payment, but what if there are several?

Suppose I told you that I put $500 into a fixed-rate bank account that paid compound interest once a year, and that after three years I had $572.68, what's the obvious question to ask? It's 'What was the interest rate?'

The calculation we need to do is to back out r from

$$572.68 = (1+r)^3 \, 500.$$

Rearranging this we get

$$\sqrt[3]{\frac{572.68}{500}} - 1 = 0.0463 = 4.63\%.$$

Generally, with a principal of P, a final sum of V, n payments every year, for T years then

$$r = n\left(\sqrt[nT]{\frac{V}{P}} - 1\right).$$

Forward rates

Recap. We've looked at simple and compound interest. We've done interest paid at different time intervals, and different number of payments. And we've also seen easy examples of backing out interest rates.

What we haven't done, yet, is look at interest rates that change depending on how long you lock away your money. For example, lock up your $100 for one year and get $104. Lock it away for two years and you get $107. I've used simple numbers so that it is very obvious that the interest rates must be different. And this is nothing to do with compounding, interest on interest, or conventions around annualizing. No, there seem to be two different interest rates in play here, one for one year and one for two years.

If I back out the annualized rate from the two-year account I get

$$\sqrt{1.07} - 1 = 3.44\%.$$

(The square root is because the maturity is two years.) The problem is that according to this calculation the rate of 3.44% is being used for both years of the second fixed-rate account. How can we have a rate of 4% for one year and *simultaneously* a rate of 3.44%? This would be fine if for example the bank holding our money for the one year was at risk of default, and so had to give a higher rate to attract investors than the bank holding our money for two years. But what if there is no risk no default or if the two accounts are with the same bank?

There is a simple solution to this problem, and that is to introduce interest rates that vary with time. We will have one interest rate for the first year, and a different interest rate for the second year. Let's see how this works with our numbers, and some symbols.

Look at the first bond to mature. It's clear that the interest rate must be 4%, not much mathematics required there since it's one payment and after one year.

Ok, so let's work with an interest rate of 4% for the first year, *even for the second fixed-rate account*. And call this rate r_1.

Our $100 in the two-year account becomes $104 after one year. Now we ask, 'What rate must apply to that $104 over the second year to give us $107?' Call it r_2.

That's

$$(1 + r_2)\,104 = 107.$$

And so

$$r_2 = \frac{107}{104} - 1 = 2.88\%.$$

How's that? For one year the interest rate is 4%, then for the second year the interest rate falls to 2.88%. And that's a *forward rate*, one that applies between two times both in the future. In this case, the 2.88% applies from the start of the second year right up to the end of the second year.

In symbols, if we use V_i to mean the value of our initial money plus *annual* interest payments after i years and r_i to be the interest rate over the ith year then we have

$$r_i = \frac{V_i}{V_{i-1}} - 1,$$

with V_0 being the initial investment.

With n interest payments a year you'll need to adjust this.

Now you are starting to see how fixed-income analysts think!

Floating interest rates

Floating interest rates are very similar to continuous compounding but with a twist, the rate r can change at any time, at the whim of the bank. (Ok, not really at their whim. If they dropped their rates too much then people would take their money elsewhere!)

We saw some of the mathematics of varying interest rates when we looked at forward rates, but with floating rates we don't know what those rates will be in advance. Very often we assume that those interest rates are random. And that's a whole new ballgame. There's a little bit about that new ballgame in the Appendix.

What Next?

If you continue with mathematical finance at university then you'll be learning about more complicated fixed-income instruments and using more advanced mathematical techniques.

The other financial products you'll be seeing include swaps, bond options and various instruments that depend on whether or not a company defaults on interest payments or goes bankrupt.

Swaps are contracts between two parties in which they exchange, or swap, cashflows from other simpler instruments. For example, a swap might be a contract to exchange fixed interest payments for floating payments. Bond options are contracts giving the owner the right, but not the obligation, to buy a bond at some time in the future. (When you've read the chapter on options that might make sense.) Crucially, valuing bond options requires a mathematical model for the random behaviour of interest rates.

In this book we are only working in what is called 'discrete time.' This means that events such as interest payments (or later, the change in share prices) happen at discrete time intervals. University-level financial mathematics tends to be in 'continuous time' in which interest rates potentially change constantly. Because of this, the kind of mathematics you'll be using includes 'calculus' and 'ordinary differential equations.' In the Appendix there is a little bit about continuous time because I give the example of what happens in the limit as the size of interest payments tend to zero, while the number of them tends to infinity!

Exercises

1. With a rate of 4.3% paid annually, what does $100 become after one year, two years, and five years if interest is compounded?
2. $100 becomes $111.35 after three years. If interest is paid once a year at the same rate, what would that rate be under a) simple interest and b) compound interest?
3. The annual interest rate is 6.2%, paid every six months. Under compound interest how much does $250 become after four years? (Check the convention around annualizing interest rates.)
4. An annualized interest rate of r is paid, compounded, n times a year. What is the general formula for the total interest paid after one year? Expand this formula for n being 2, 3 and 4.
5. The annual interest rate is 5.9%, paid every three months. Under compound interest how much does $150 become after 18 months?
6. The following table shows how much $100 in a floating-rate bank account becomes after various time periods. Calculate the annual forward rates. The first few have been done for you. (For clarity, the 2% on the right refers to the rate from zero to one year, the 2.55% refers to the rate from one to two years, 3.35% from two to three years, and so on.)

Maturity	Total value	Forward rate
0	100	N/A
1	102	2.00%
2	104.6	2.55%
3	108.1	3.35%
4	111.4	
5	116.1	
6	120.7	
7	125.9	
8	131.5	
9	136.8	
10	142.6	

7. Which gives the better return when compounded, 5% p.a. for two years or zero for one year followed by 10% for one year? How much does $100 become in the two cases?
8. A floating bank account pays 2% for one year, 4% for a second year and then 6% for a third year. How much would $100 become with these rates compounded? What fixed interest rate compounded annually for the three years would return the same amount?

IN DEPTH: SHARES

Shares are part ownership in a business. It could be that your family owns 50% of a fruit and veg store. But that's not usually what we mean. Shares in small businesses are difficult to value, they don't change hands very often, and are highly illiquid. If you need to get out of the family fruit and veg business you can't just call up a stockbroker.

In the subject of mathematical finance, we usually mean shares in a large company, such as Apple. In which case you wouldn't be owning 50%. You'd be lucky to own 0.00000003% of Apple. But on the plus side, if you needed to sell your shares in Apple it would be easy, it's a very liquid stock, meaning that there are always people wanting to buy or sell Apple shares. And because those shares are publicly traded, we know their value.

The most interesting mathematical model of share prices is to treat them as random. The idea is that prices bounce up and down all the time, in an unpredictable way. Now whether prices are really random is open to debate, but it's certainly true that such models for share prices, assuming randomness, have been phenomenally successful in building theoretical models for more complex financial instruments, and in asset management.

But what exactly do we mean when we say that share prices are random? It means that we might as well pretend that share prices from one day to the next are determined by the toss of a coin, heads and the stock rises, tails it falls. Or we have a hat full of random numbers, pull one out, that tells us how much the stock value changes.

If we are playing a game of backgammon then we also assume that the result of the roll of the dice is random and play accordingly, that's even though there is physics behind the roll. But that physics is very, very, very complicated. It's similar with share prices, maybe there is a mathematical model that could be used for predicting share prices exactly, but it would also be prohibitively complicated. Just like the dice roll where our backgammon play assumes randomness, with shares we assume randomness. Although we can't predict the future, we can use such models to tell us the probabilities of future share price movements and then invest accordingly.

But we don't just pick one random number and model tomorrow's share price. No, we pick many random numbers, so that we get a random simulation of the stock price from today, to tomorrow, then to the day after, and the day after that, and so on, as far into the future as we want.

However, the model needs to be realistic. We can't just pick any old random numbers. The model needs to be based on the past behaviour of the stock price. The starting point for modelling share prices has to be a combination of an analysis of the statistics of the stock price in the past, that's historical data, and a bit of common sense.

The Mathematics

Looking at returns

Here's a plot of the Apple share price over about a year.

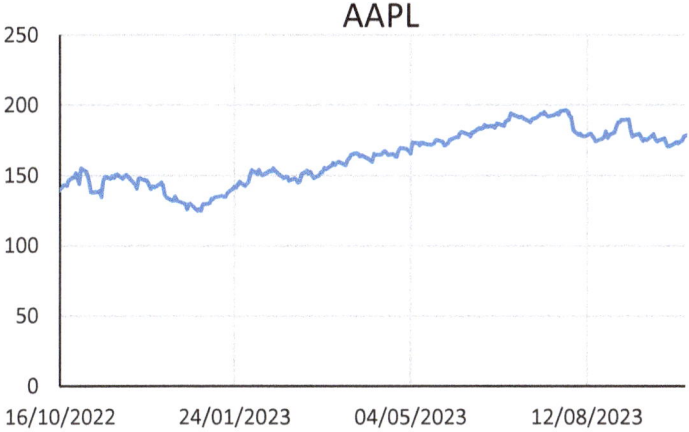

This data is readily available for free (not all financial data is free!) on https://finance.yahoo.com/quote/AAPL. AAPL is the ticker symbol for Apple. On that website look for the link to Historical Data.

How should we analyze this data to get clues to the modelling as random? Well, here's an observation. It might at first sound counterintuitive. But it's an important insight into financial modelling. Ready?

As an investor, we don't care about the share price!

[Editor: 'Surely there's some mistake, Paul.']

No, let me say it again, as an investor we don't care about the share price.

What we care about is by how much the share price goes up (or down).

Ok, maybe I'm being a little bit naughty in saying we don't care about the share price. But think of it this way, it doesn't matter whether a share price is $1 or $100, because if we want to invest $500, it just means that we have to either buy 500 or five shares.

What we then care about, having put $500 into this stock, is the *return on our investment*. In other words, what is the percentage change in the share price over some timeframe.

And that's what we are going to analyze, the day-to-day share price return.

Here are some of the numbers that made up that AAPL plot, presented in a spreadsheet.

	A	B	C	D	E
1	Date	AAPL	Return		
2	28/09/2023	170.69			
3	29/09/2023	171.21	0.0030	=(B3-B2)/B2	
4	02/10/2023	173.75	0.0148		
5	03/10/2023	172.40	-0.0078		
6	04/10/2023	173.66	0.0073		
7	05/10/2023	174.91	0.0072		
8	06/10/2023	177.49	0.0148	=(B8-B7)/B7	
9	09/10/2023	178.99	0.0085		
10	10/10/2023	178.39	-0.0034		
11	11/10/2023	179.80	0.0079		

Column *A* is the date. Column *B* is the share price at the time the market closed. Column *C* is the return from one day to the next. It is easily calculated. In Cell *C3* the return is *B3* − *B2* divided by the share price on the first day, *B2* again. Note that this return is over each day, it is not the return as measured from some start date.

And here is a plot of all those returns, for the whole year.

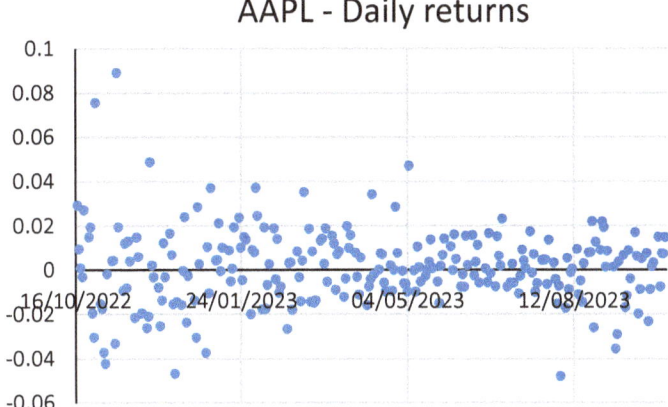

I haven't joined the dots in this particular plot because we are going to be assuming that the return on one day is independent of returns on previous days, so there's no connection between them, and so no reason to join the dots.

Mean and standard deviation of returns

It's easy enough to calculate the mean and standard deviation of returns from this data, in Excel there are inbuilt functions that will do this for you. And this is what I found.

$$\text{Mean:} \quad 0.001217 = 0.1217\%$$
$$\text{Standard Deviation:} \quad 0.01708 = 1.708\%$$

The mean is positive, so the Apple share price has been rising. But there's a lot more randomness, measured by the standard deviation of returns, than growth, measured by the mean return.

Ok, so we've now got some statistics for how AAPL has behaved over one year. What's a good model for this randomness going forward. Or, to start with, what is the *simplest* model?

Symbol time!

The return each day is given by

$$R_i = \frac{S_i - S_{i-1}}{S_{i-1}}$$

where S_i is the share price on the ith day.

Suppose we know the share price, yesterday, that's S_{i-1}, and the return R_i, then we can rearrange this as

$$S_i = (1 + R_i)S_{i-1}.$$

This gives us the important modelling clue, that the next share price is the current share price *multiplied* by a factor. Hold that thought.

Now back to the question, what is the simplest model? One that has the multiplicative behaviour that we want.

The simplest model

The simplest model has to be a coin toss, there's no more basic model of randomness. Schematically this would look like this:

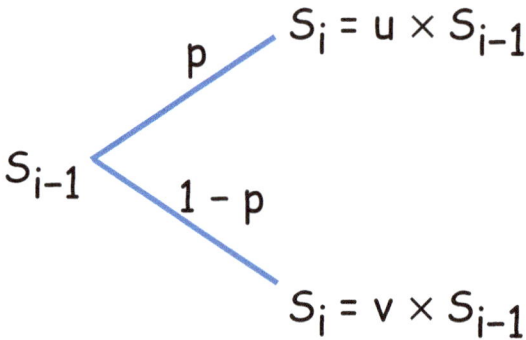

This schematic is understood as follows. The share price now is S_{i-1}. Tomorrow it is S_i, after rising or falling. If it rises then its value is uS_{i-1} or if it falls vS_{i-1}. Here $u > 1$ and $v < 1$. It rises with a probability p, or falls with a probability of $1 - p$. (The probabilities have to add up to one.) I'm allowing for the coin to be biased, p doesn't have to be 0.5.

This is called the binomial model.

This model has three parameters that we need to choose, the u, v and p. And we need to choose them so that this model has the properties that the mean return is 0.1217% and the standard deviation is 1.708%.

The expected return from the model is

$$\frac{puS_{i-1} + (1-p)vS_{i-1} - S_{i-1}}{S_{i-1}} = pu + (1-p)v - 1 = 0.001217.$$

That's our first equation.

The variance of returns is

$$p(u - pu - (1-p)v)^2 + (1-p)(v - pu - (1-p)v)^2$$
$$= p(1-p)(u-v)^2.$$

(Try that yourself. It's easier than it looks!)

Our second equation comes from the standard deviation, the square root of this,

$$(u-v)\sqrt{p(1-p)} = 0.01708.$$

Interestingly, these are *two* equations (one for the expected return and one for the standard deviation) for the *three* unknowns (the u, v and p). Usually you'd expect to have two equations if you have two unknowns. Or three equations for three unknowns. What we have here is not enough information to pin down our three unknowns. Or to put this another way, in a more positive light, there is a lot of freedom in how we choose our three parameters. There is an infinite number of solutions that work, so I'm going to find one solution that is particularly nice.

I'm going to make this really simple for now — later I will tweak this — and fix the probabilities at 0.5, p = 0.5. In other words, let's work with a fair coin. In this case, we solve these two equations to get

$$u = 1.0183 \text{ and } v = 0.9841.$$

So that's our simple model. Toss an unbiased coin, heads and the share return is 1.83%, tails and it's -1.59%.

Simulations

Let's start off with the stock price at $179.8, as that's the last share price in my data, let's pretend it's today. We toss the coin. It is heads. The share price becomes 1.0183 x 179.8 = 183.1. After one day this is our share price plot.

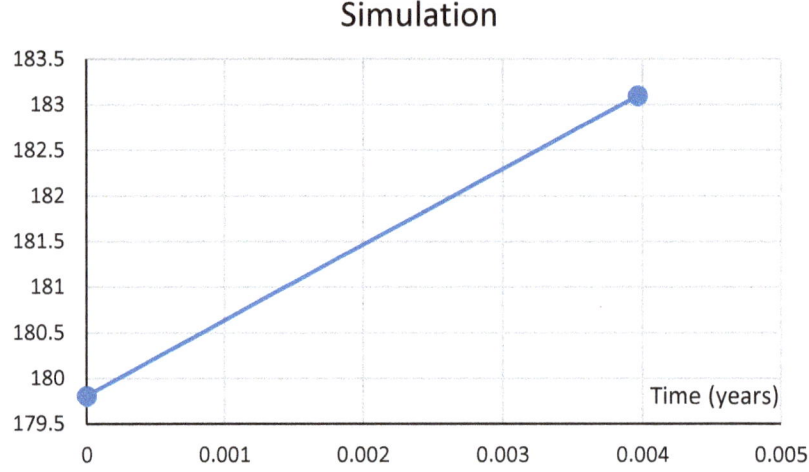

The next day we toss heads again, the share price is then 1.0183 x 183.1 = 186.8, and after that it's tails, and the share price is 0.9841 x 186.8 = 183.5. Here's the simulation after three days/tosses.

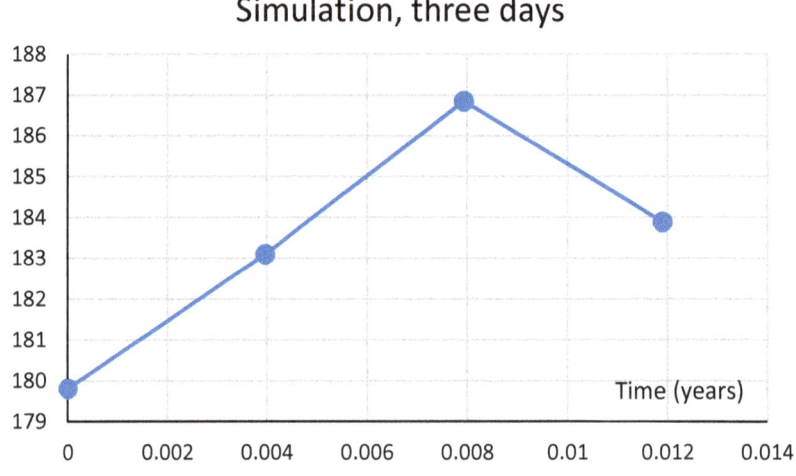

I don't know whether you've noticed, but the horizontal axis is time measured in years. And one business day is approximately 1/252 of a year. (Markets aren't usually open at weekends.) Hence the small numbers along the horizontal axis. Now let's toss that coin 252 times to get a simulation of the AAPL share price over one year.

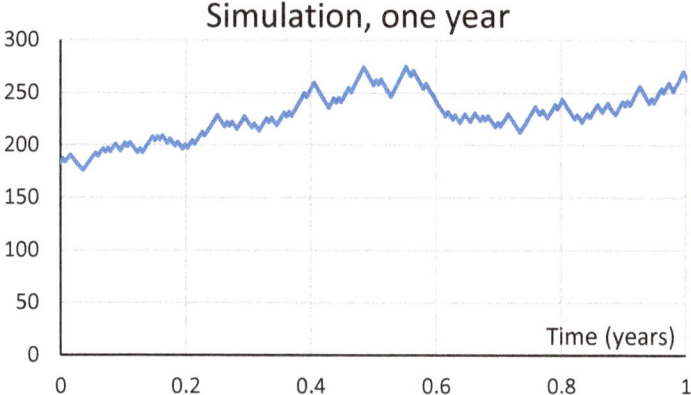

This is a simulation of what might happen to the Apple share price over one year. It is most definitely *not* a prediction! A single simulation like this isn't much use, as it's never going to happen. Instead, what we do in practice is to do many, many such simulations over the same period. The above is one simulation, now we go back to time zero, share price of 179.8, and toss another 252 times. That's the second simulation. Repeat. And repeat. Maybe repeat a thousand times or more! We end up with a picture like this.

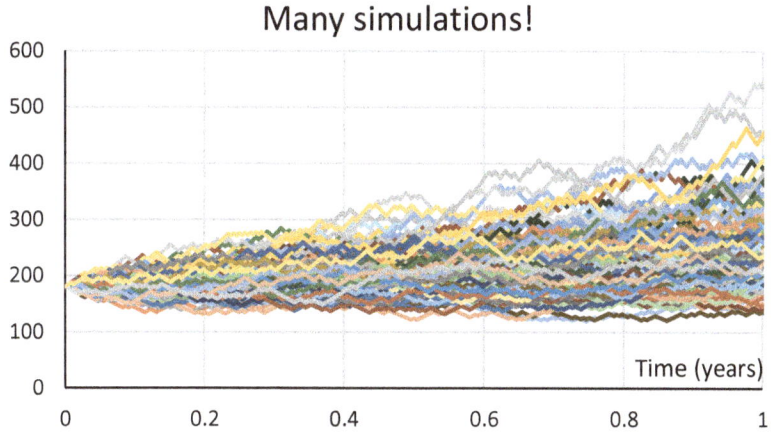

Each of those coloured lines is one realization of the possible share price path over one year. We can use simulations like this to answer questions such as, 'What is the expected share price after one year?' Just calculate the average share price at year one from all the simulations. The simulations tell us that the expected share price is $244. That's a pretty good expected return then.

Or, 'What is the probability of the share price being above $179.8 in one year?' Easy. We simulate share price paths many times, count how many simulations have the share price above $179.8 after one year, and divide by the total number of simulations. I've done this for these paths and I find that there is an 86% probability of the Apple share price being above $179.8 in one year. That's useful information, it tells us how likely we are to make a profit. But really we hope that the share price is not just more than we bought it for but more than the initial price plus interest we would have received if we'd put the money in the bank. Supposing an interest rate of 5%, my simulations tell me that there is an 82% chance of the share price being above $188.8 in one year.

We could get more sophisticated than this. Instead of waiting to see what the share price is after one year we could decide that we sell as soon as the share price hits $215, we don't want to be too greedy, or take more risk than we need, after all the share price might rise to start with and then fall. It turns out that the probability of the share price ever reaching $215 in one year is 75%.

All those numbers are pretty good, it looks like AAPL is a very good investment. But here's a warning for you, those results all hinge on the parameters we had for the mean and the standard deviation. And those parameters came from the data. But the data I chose was just one year in the life of the Apple stock. Had I chosen a different period I would have got different results. This illustrates that it is almost always the case that you need to use as much data as possible, and over a long period of time.

And another warning, I didn't simulate enough paths for accurate probabilities. I only did just over 100 paths to get the average of $244, the probability of 86%, etc. For accurate results you need at least 10,000 paths! The more paths, the more accurate the results.

Arithmetic and geometric random walks

The model for the share price that we've built up here is an example of a geometric random walk. It's slightly more advanced than an arithmetic random walk. What are these two random walks? Indeed, what is a random walk?

A random walk is the path you get by moving up, down, left, right, randomly, perhaps according to some mathematical model. It could be in three dimensions, with a balloon being bounced around by the wind. It could be in two dimensions, a puck being knocked around during an ice hockey match. But here it's a pencil on graph paper going up or down at each timestep according to the toss of a coin.

The arithmetic random walk is one in which you add or subtract a number at each time step. And it's the same numbers every time. Something like, 100, to 101, to 102, to 101, to 100, 99, 98, 99, 98, 97, 96, 95, 96, and so on. This is very simple, but suffers from one technical problem if it's meant to represent a share price ... there's nothing stopping the numbers going negative! And that's not a great model for a share price. Think of it like this: Start with a value of 100, add or subtract one depending on whether you toss a head or tail. Keep tossing the coin. Eventually the number might reach 100 + 200 = 300, that's easy to imagine. But then 100 - 200 = -100 is equally likely.

The geometric random walk is also very simple, but involves *multiplying* by numbers each time step. Something like 100 to 101, to 102.01, to 100.9899, to 99.98, and so on. This random walk can *never* hit zero. It can get very, very close, but never reach it. And that's the random walk we are using here, thanks to our observation about modelling returns. It's the obvious choice for representing share prices.

Expected growth and volatility

I downloaded data from Yahoo! for the prices of Apple stock. The data I used was daily. But it could have been hourly. Or weekly. And so just like we had with interest rates, conventions have developed as to how to quote the properties of share price behaviour.

Our data gave us an expected return of 0.001217 and standard deviation of 0.01708 *over one day*. But someone else might be quoting expected return and standard deviation over one week. How can we make comparisons if different people are using different timeframes? We have to translate those numbers into equivalent annualized numbers.

Since expectations can be added, we can convert from a daily expected return to an annualized quantity, known as the annual growth rate, by multiplying by the number of business days in one year, 252. So we would quote an annualized growth rate of 252 x 0.001217 = 0.3074 = 30.74%.

It's a tiny bit more complicated to annualize the standard deviation because we can't simply add up standard deviations.

First, remember that the standard deviation is the square root of the variance, and that we *can* add up variances if the returns from one day to the next are independent of each other. So, to get to an annualized *variance* we multiply by the number of business days in a year. And so to get an annualized standard deviation we need to multiply the daily standard deviation by the *square root of the number of days in a year*, that's about 15.9, giving, for AAPL, 15.9 × 0.01708 = 0.2711 = 27.11%. This is called *volatility*.

In words, we say that the expected growth rate of Apple stock is 30.74% *per annum*. Just like we must say whether a length is inches or centimeters, etc. The symbol often used for growth rate is μ, the Greek letter, spelled mu in English and pronounced 'mew.'

And the volatility is 27.11%. Technically we ought to say 27.11% *per square root year*. But the convention is to not bother saying the per square root annum. I suspect no one does this because very few people know the Latin for 'square root'! The symbol used for volatility is σ, the Greek letter sigma.

Here's a plot showing a selection of stocks with their expected returns and volatilities. Even this simple plot hints at which assets look like better investments than others. For example, **A** looks good with a very high expected return, but it comes at the cost of also very high risk.

Stock **D**? Not as high an expected return as **A**, but also not as high a risk. A possible contender for someone who is very risk averse?

And **E**? Looks pretty rubbish, it has quite a low expected return, and a very high risk. Best avoided.

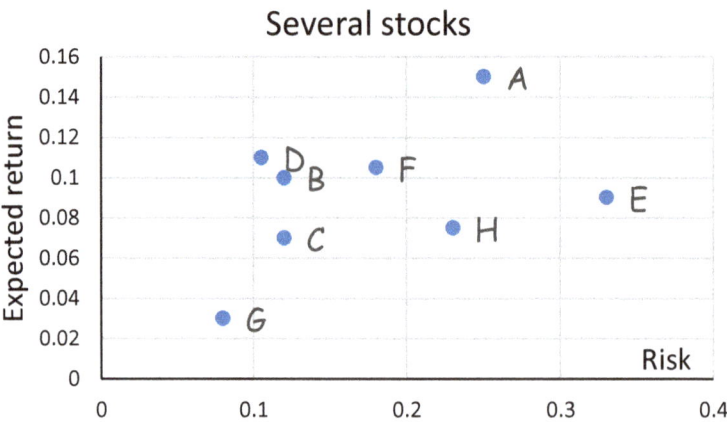

However, it's not that simple. **E** might be a poor investment on its own, but maybe it's good for selling short, so you can invest more in **A**? Or maybe it has a correlation with other stocks that makes it work well in a portfolio, correlation can be used to reduce risk. Details in the next chapter!

Sharpe Ratio

A simple way of comparing and contrasting risky investments, before we get onto something more sophisticated in the next chapter, is via a quantity known as the Sharpe Ratio, after Nobel laureate William Sharpe.

Invest in stock **A** and you will have a certain expected return, but it comes with a certain risk. However, you can get a decent return just by putting your money into a risk-free bank account. So any investment we make ought to have an expected return greater than the risk-free rate. It makes sense to look at

$$\mu_A - r.$$

Here μ_A is the expected growth of stock **A**.

This is the *expected return in excess of the risk-free rate*. But that doesn't allow for the risk we are taking. So let's work with the ratio of this excess return to the risk taken, that's just

$$\frac{\mu_A - r}{\sigma_A}$$

where σ_A is the volatility of stock **A**. This is now the *excess return per unit of risk* or *risk adjusted expected return in excess of the risk-free rate*. It's known as the Sharpe Ratio. It's a clever way of comparing and contrasting different risky investments. This ratio increases as the expected return increases and the risk decreases. We can see what this means with the following plot.

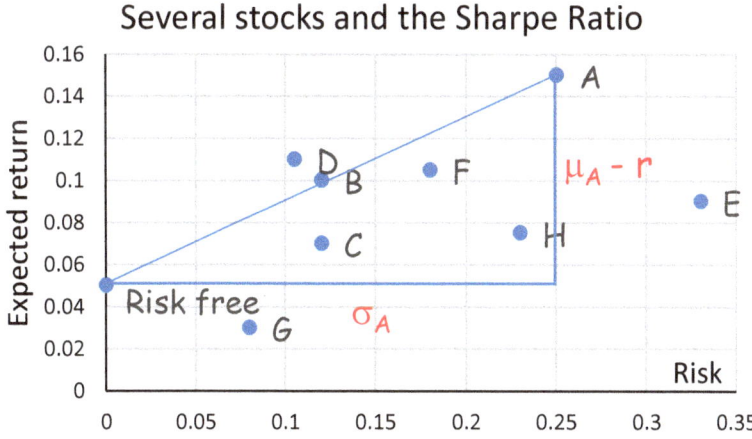

You can see the numerator and denominator for stock **A** in red. The Sharpe Ratio is then just the slope of the line joining the risk-free investment and the stock. The higher the Sharpe Ratio, the steeper the line, and the better the stock is as an investment.

Yes, the Sharpe Ratio is a great way of comparing individual risk investments. But it doesn't take into account how investments behave when there are several stocks, different companies, in the one portfolio. And that's coming next.

What Next?

As with fixed-income mathematics, when you get to university you'll be modelling share prices in continuous time. No more having share prices changing value every day, they change continuously, faster than every femtosecond! This entails using a kind of mathematics called 'stochastic calculus.' And this in turn leads on to 'partial differential equations.'

You will still work with expected share-price growth and its volatility. But if you get to analyze share-price data you will quickly realize two things. First, it's very difficult to figure out what the expected growth is. And, second, while volatility is statistically easier to measure, it will become clear that it changes. It's not the nice constant we've had here. So at university expect a lot more statistics!

Exercises

1. A stock price now is $76.76. In six months it is $84.67. What was the return over that period? Convert that to an annualized rate. Explain the convention(s) you have used.
2. A stock price is now $43.55. In three months it is $39.98. What was the return over that period? If the return is the same for the next three months, what would the price be then?
3. A stock price is now $124.35. After one year it falls to $110.96. What is that return? How much must the return be over the next year to get the price back up to $124.35?
4. The share price today is $141.1. Tomorrow it could rise to $145.2 with probability 61% or fall to $138.9. What is the expected return over the day, and the standard deviation of the return?
5. The share price today is $141.1. Tomorrow it could be the same with probability 29%, rise to $145.2 with probability 35% or fall to $138.9. What is the expected return over the day, and the standard deviation of the return?
6. A share price starts at $100. It is equally like to go up or down by 1% every day. After one day is the average price higher or lower than $100, or exactly $100? And after two days? Higher, lower or the same? And after one year? What do you notice about the position of the mode? Is the distribution of future stock prices symmetrical or skewed?
7. A share price starts at $76.80. Every day, there is a 60% chance of it rising by 1.2%, and 40% chance of falling by 0.8%. What is the probability of the stock being above $81 after one week (of five business days)?

8. Convert the following average returns and standard deviations to annualized quantities. The first two have been done for you.

	Period	Average Return	Standard deviation	Expected growth (annualized)	Volatility
A	1 day	0.0496%	1.241%	12.50%	19.70%
B	1 day	0.0103%	0.881%	2.60%	13.99%
C	1 week	-0.4144%	2.356%		
D	2 days	0.0452%	0.803%		
E	2 weeks	0.3788%	4.462%		
F	1 month	-0.7500%	7.280%		
G	1 week	0.3490%	3.893%		
H	2 weeks	1.6269%	6.258%		

9. What are the Sharpe Ratios for the following stocks? The risk-free interest rate is 4.7%. The first two have been done for you.

	mu	sigma	Sharpe Ratio
A	0.15	0.25	0.412
B	0.1	0.12	0.442
C	0.07	0.12	
D	0.11	0.105	
E	0.09	0.33	
F	0.105	0.18	
G	0.03	0.08	
H	0.075	0.23	

IN DEPTH: PORTFOLIOS

You have a portfolio when you have more than one investment. It might be a portfolio of risk-free bonds, risk free because you know exactly what return you will be getting. Or a portfolio of risky shares, that would be risky because you wouldn't know for sure what return you would get. Most investors though would have a portfolio that is a combination of both risk free and risky.

If you own shares then you are exposed to the risk of them falling in value. (I'm assuming that you are long only, that you haven't sold any stocks short. If you are short stocks you will be worried about them rising in value!) If you have many companies in your portfolio then it is unlikely that all will fall simultaneously. Yes, some may fall, but others will rise. This is the idea that you shouldn't put all of your eggs in one basket. This diversification can work really well, but it does depend on how correlated the assets are. If there is a high degree of positive correlation between the assets then there is also a high chance they'll fall together. So, how well you are diversified depends on the correlations between the shares in your portfolio.

Measuring correlations is easy enough, I've described the formula already. But there is one big problem here. Correlations are notoriously unstable. More later, along with a brief discussion of stock market crashes!

The Mathematics

We've done the mathematics of fixed-income investments. And we've done the mathematics of individual shares. What if we have a portfolio of one bond and one share? This will be our starting point for the analysis of portfolios.

One bond and one share

We'll have a whopping total of $1 to invest! Seems like not very much, I know. But everything we do is going to be linear in the amount invested. If you really have $1,000 to invest then just multiply the relevant numbers by 1,000. So, all I'm doing now is a mathematical trick that reduces the number of numbers or symbols I need, it keeps things uncluttered.

Ok, we have $1 to invest. And I'm going to put b of that into the stock and the remaining 1 - b into the bond. For example, b might be $0.50, that would be an equal distribution of our initial cash.

The bond has an annual interest rate of r.

The share has a current price S_0, annualized expected growth rate of μ and volatility (annualized standard deviation of returns) of σ. Because we are buying b *worth* of shares that means that we must buy a *quantity* b/S_0 of them.

Let's look at what the portfolio value will be after one year, say. In fact, in all of my portfolio examples I'm going to make things simple by assuming we buy the constituents of the portfolio and then don't look at the portfolio value for one year. We could get more complicated by looking at our portfolio every day and then rebalancing by buying or selling depending on the performance up to that date, but that's university-level mathematics! Another reason for looking at the portfolio after one year is that we don't have to worry about interest rate conventions, and so on.

After one year the money in the bond becomes

$$(1 - b)(1 + r).$$

That's the initial bond investment plus the interest.

The money in the share becomes the number of shares we bought multiplied by the now share price:

$$\frac{b}{S_0} S.$$

Here S is whatever the share prices is after one year. But in terms of the random share price *return*, R, we have

$$S = (1 + R)S_0.$$

And therefore the money in the stock is

$$b(1 + R).$$

We can calculate the *change in value* of our portfolio of bond plus shares after one year, and because we started with $1 this is the same as the *return*. It is

$$(1 - b)(1 + r) + b(1 + R) - 1.$$

In this we know b, since we decided how much we put in each of the bond and shares, and we know r, the risk-free interest rate. But we don't know the random return, R. At least not now, when we set up the portfolio. We will know the return after one year, but by then it's too late to change our portfolio!

We can easily calculate the probabilistic properties of this expression.

The mean of this is just

$$(1 - b)(1 + r) + b(1 + \mu) - 1 = (1 - b)r + b\mu.$$

The variance is

$$b^2 \sigma^2.$$

So that the standard deviation of returns is

$$b\sigma.$$

Before we look at what these two expressions mean, I want to describe a little bit of financial economic theory. 'Theory' might not be the right word. Maybe 'assumption' is better. No, let's call it a little bit of financial economic 'hope.' It's simply the idea that the return on risky stocks jolly well ought to be greater than the risk-free interest rate. Why? Because if the expected return on a stock is less than the risk-free interest rate why would anyone invest in it? Not only would you expect to get less than putting money in a safe investment, the bond, but there's also a chance that the stock will fall even further!

This is the idea that people should be risk averse, discussed earlier. So in the following examples I am going to assume that not only is μ greater than zero but also that it is greater than the risk-free rate, so $\mu > r$.

Now let's look at a plot representing the mean return and its standard deviation,

$$\text{Expected return} = (1 - b)r + b\mu$$

$$\text{Standard deviation of returns} = b\sigma.$$

In the plot below, the horizontal axis is the standard deviation, the risk, and the vertical is the expected return. The dot at the bottom left represents the risk and return for the risk-free asset. It's exactly on the vertical axis since the risk is zero. The other dot, labelled **A**, shows the risk and expected return for our stock. I draw the line by picking a value for **b**, this gives me a value for each of the standard deviation and the expectation, and that becomes a single dot. I then vary that value for **b** between zero and one, and that gives me a line of dots … well, just a line. Then I also look at what happens if **b** is greater than one. This means that I have borrowed money at the interest rate of **r** to invest in the stock, that's the thinner line in the plot.

In this picture I have a risk-free rate of 5%, an expected growth rate for the stock of 15% and a volatility of 25%.

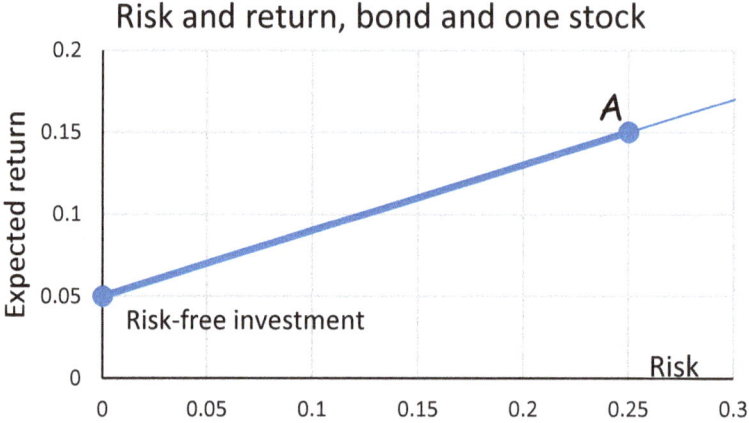

This plot shows beautifully how the more risk you take the more you expect to be rewarded. Of course, this is contingent upon the assumptions, that we know the parameters, and that the expected return on the stock is greater than the risk-free rate. If the expected return was less than the risk-free rate then the line would have a negative slope, it would slope downwards going to the right.

You can also see that if you calculate the Sharpe Ratio for any point on this line it will be the same. In order words, the expected return on the portfolio in excess of the risk-free rate after being adjusted for risk is always going to be the same as the stock on its own.

If you drew a plot like this using real data from your favourite stock, can it tell you how much to put into the stock? How large an investment would you personally make? Probably not. Classical financial economics tells us that we need to know something about the investor before we can say where on this line they should be, what value of b to choose. If you are extremely risk averse, you'd want to be near the left-hand side of this plot. With $b = 0$ you have zero risk and receive the risk-free rate of return. If you are comfortable with taking financial risks then you'd be further right. More shortly.

Let's now look at the mathematics of investing in two different, but correlated stocks. But no risk-free asset, yet. That's a bit later.

Shares in two companies, no risk-free investment

We'll have two shares, S_A and S_B. And all other notation is similar to that above, and subscripts on the parameters have the obvious meanings. We'll invest an amount b in the first stock and $1 - b$ in the second. I can go straight to writing down the change in the portfolio value after one year, in terms of the random returns on the two stocks:

$$b(1 + R_A) + (1 - b)(1 + R_B) - 1 = bR_A + (1 - b)R_B.$$

The expected growth of this portfolio, μ_P, is very easy to calculate. It is

$$\mu_P = b\mu_A + (1 - b)\mu_B.$$

The standard deviation is harder.

The variance is given by

$$E\left[\left(b(R_A - \mu_A) + (1 - b)(R_B - \mu_B)\right)^2\right].$$

This becomes

$$b^2 \sigma_A^2 + 2b(1 - b)E[(R_A - \mu_A)(R_B - \mu_B)] + (1 - b)^2 \sigma_B^2.$$

You'll recognize the terms at the two ends, they are just the same as the variance when we had a single stock. The term in the middle is new. And this is the term representing the contribution to the variance of the correlation between our two risky assets. You might recognize it from the chapter of mathematics. With the symbol ρ meaning the correlation between the returns on the two stocks, a number from minus one to plus one, the variance of the portfolio is

$$\sigma_P^2 = b^2 \sigma_A^2 + 2b(1-b)\rho\sigma_A\sigma_B + (1-b)^2\sigma_B^2.$$

And the standard deviation is the square root of this.

We can do a similar plot of expected return against risk, the standard deviation, as before. As above, for each choice of b we get a different coordinate, risk and expected return, and varying b gives us a line. The difference between our new plot and the previous is that we no longer have a straight line, we have a hyperbola.

In my plot, stock **A** has the same parameters as before. But now the expected growth of stock **B** is 10% with a volatility of 15%. The correlation between the two stocks is 20%, 0.2.

Ignore the colours in this plot for a second. Again, the bold line is what can be achieved with b between zero and one. With b less than zero or greater than one we have to sell one of the shares short to invest more in the other. That gets us onto the thin lines.

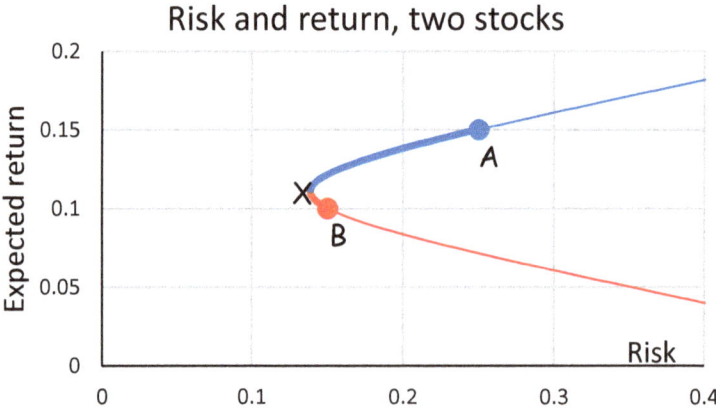

Now look at the colours. The top half of this hyperbola is blue and the bottom half is red. Red, if not for danger, at least for 'to be avoided'! Suppose we decide we are comfortable with a risk, a standard deviation, of 0.2. We can get this with two portfolios. There's one on the red line and one on the blue. Just go to the value 0.2 on the horizontal axis and move upwards until you reach the hyperbola.

These are two different portfolios, one is long both shares (the one on the bold blue part of the line) and the other requires going short one of the shares (it's on the thin red line). But there's another difference between these two portfolios, and this is of profound importance. One has a return of about 14% and the other about 8%. Which portfolio would you prefer?

Obviously, you'd rather have the portfolio that's on the blue part of the line. That portfolio has the same risk as the one below on the red line but has a much higher expected return.

The 'efficient frontier' is made up of the portfolios that give the highest expected return for a given level of risk (or the lowest risk for a given level of expected return). And that's the blue part of the line. The X in the plot marks the point at which the portfolios flip from efficient to inefficient. The A and B points are the two shares making up the portfolio.

Where you should be on the efficient frontier, how to decide the value of b, is a personal decision depending on how risk averse, or not, you are.

Shares in many companies, no risk-free investment

There's not much mathematical difference between a portfolio with shares in two companies and a portfolio with shares in many companies. If I use b_i as the amount invested in the ith company, with similar meanings for μ_i, etc. then we have an expected return of

$$\sum_{i=1}^{n} b_i \mu_i$$

with n being the total number of companies. Obviously, the bs must all add up to one again. The variance of the portfolio is

$$\sum_{i=1}^{n}\sum_{j=1}^{n} b_i b_j \rho_{ij} \sigma_i \sigma_j$$

with ρ_{ij} being the correlation between the ith and jth stocks with the understanding that if $i = j$ then the correlation is one.

The expected return/risk plot is similar to what we saw with just two stocks.

But when we add in the risk-free investment something special happens!

Shares in many companies, with a risk-free investment

Now include the risk-free investment. While the expected return changes, the risk doesn't since the risk-free investment is, and there's no better way to say this, risk free. The expected return is

$$\left(1 - \sum_{i=1}^{n} b_i\right) r + \sum_{i=1}^{n} b_i \mu_i .$$

Now the **b**s don't have to add up to one, the balance (the amount in front of the **r**) is put into the risk-free investment. We can draw the following plot.

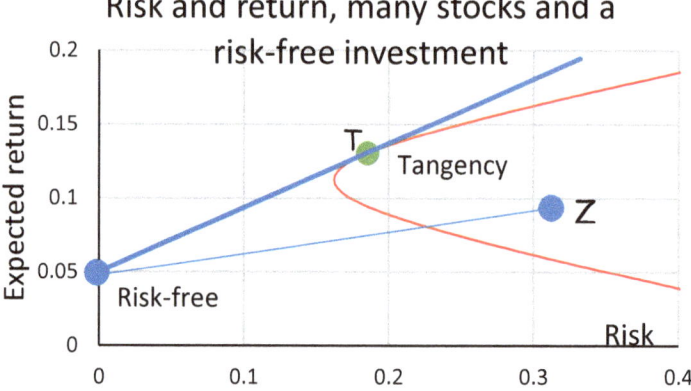

This plot is a combination of the two previous plots. The red curve represents the possible risk and returns you can get with a portfolio of stocks without the risk-free asset. However, because we now have many different stocks, instead of just the two, it is possible to have portfolios with risk and expected return *inside* the red curve, not only along it. The point Z represents one such possible portfolio.

Then when you add the risk-free investment to the Z portfolio you get a straight line going through the left-hand dot and through Z, as shown. But as illustrated, that is nowhere near the best you can do in terms of risk versus return. You can do much better. Just imagine rotating this blue line anticlockwise so that it passes through the same risk-free, dot, and through other possible portfolios other than the Z.

The best possible portfolios are then those that are on the bold blue line in the picture, they have the highest Sharpe Ratio. Those are the investments that have the highest possible expected return for any given risk, they are efficient. And they are made up of a combination of the risk-free asset and the very special portfolio of shares, marked T in the plot, known as the Tangency Portfolio. And again, it's a personal choice where on this line you should be.

Market crashes

A serious problem with Modern Portfolio Theory is that correlations are unstable. Perhaps you have two oil companies in your portfolio and both are equally affected by the changing price of oil. So you see a high degree of correlation between their share prices. Then one of those companies discovers a new oil field and their price rises at the expense of the other, causing the previous high degree of correlation to disappear.

In constructing a portfolio, it can be good to have assets that are uncorrelated as it decreases risk. But every now and then we see market crashes, possibly caused by some dramatic global event or by a panic, and suddenly all stocks move together … and down. We see such market crashes happen every now and then. They can be very dramatic, perhaps as much as 20% value being lost in one day. At such times diversification is not possible.

However, there are some financial instruments that increase in value when shares fall, we'll be seeing some in the next chapter.

What Next?

Things to look out for at university are: Utility functions; Continuous time, again; Consumption; And, if you are lucky, Kelly investing. I said earlier, several times, that where on the efficient frontier you should be, how you balance risk and expected return, is a personal choice. This can be modelled by a person's utility function. A utility function is a way of mathematically quantifying preferences among a choice of investments. On a university course, you might also look at the mathematics of deciding not only what amount to invest and where but also how much to *consume*, i.e. spend.

And then there's Kelly investing. This is a theory concerning how much to invest if you get many opportunities to make the same investment over and over again. Ok, it's about gambling. But is that really any different from investing?! A good university course will cover this too.

Exercises

1. The risk-free interest rate is 3.3%. A stock has annualized growth of 12.3% and volatility of 18.9%. What is the stock's Sharpe Ratio? If you split your money equally between the risk-free investment and the stock, what is the expected growth of this portfolio? And its standard deviation? And its Sharpe Ratio?
2. Same data as in Question 1. If you put b into the stock and $1 - b$ in the risk-free investment, what is the Sharpe Ratio of this portfolio?
3. Same data as Question 1. If your appetite for risk is limited to a risk or standard deviation of 10%, how much would you put in the risk-free investment and how much in the stock?
4. Same data as Question 1. If your risk appetite is 25% what would your portfolio look like?
5. You have a portfolio consisting of two stocks. The expected return of the portfolio is

$$\mu_P = b\mu_A + (1 - b)\mu_B.$$

And the variance is

$$\sigma_P^2 = b^2 \sigma_A^2 + 2b(1 - b)\rho\sigma_A \sigma_B + (1 - b)^2 \sigma_B^2.$$

Now rearrange the first of these two expressions to write b in terms of μ_P. And substitute this into the second expression. You will now have one formula for the portfolio variance in terms of the portfolio expected return. Do you recognize this as the formula for a hyperbola?

IN DEPTH: OPTIONS

You think Modern Portfolio Theory and efficient frontiers are clever? Well, now we have some more Nobel Prize winning stuff. And this is arguably even more important than MPT, and it doesn't suffer from quite as many problems, there are fewer unstable parameters! However, ...

This chapter is not easy. Well, the mathematics is easy, easier than in some of the previous chapters. But the concepts are potentially confusing. Things that seem obvious ways to value financial instruments turn out to be wrong. But not completely wrong in a way that becomes clear, but annoyingly they are only subtly wrong. In mathematical finance this topic is famously difficult to teach. People taking university degrees in this subject frequently get confused. (I sometimes wonder whether the lecturers also don't understand it!)

If you come away from this chapter thinking, 'That was easy, I don't know what all the fuss was about,' then you have probably misunderstood.

If you come away from this chapter thinking, 'That was tough, but I think I got all of the exercises correct,' then you are probably fine.

If you come away thinking, 'Frankly, my dear, I don't have a clue,' then you really shouldn't worry. Knowing that you don't know something is far better than believing you know something when you don't.

Please fasten your seatbelts and secure all baggage underneath your seat or in the overhead compartments. We also ask that your seats and table trays are in the upright position for take-off. Cabin crew, doors to manual and cross check.

Options are the right to buy or sell an asset at a time in the future for an agreed price. Specifically, the right to *buy* is a *call* option, and the right to *sell* is a *put* option.

I'll go through that again but add in a few details. And we'll only work with call options for the moment.

A call option is the right, but not the obligation, to buy a specified share at a specified time in the future for a specified price. An example would be:

> You can buy one share in AAPL for $195 on March 22nd, 2024.

The details of the option contract tell you which stock (here Apple), what price (here $195) and when (here March 22nd, 2024). (If you are reading this after March 22nd, 2024, just replace the 2024 with some year in the future.)

We need more jargon. The stock in question, the one you can buy, is called the 'underlying asset,' or simply the 'underlying.' The price at which you can buy the stock is called the 'exercise' or 'strike price.' And the date at which you can exercise this option is the 'expiration date' or just 'expiration' or 'expiry.'

This financial contract has some value. At the time of writing, AAPL is at $170. So why would a contract allowing you to pay $195 for something worth just $170 have any value? Well, expiration is five months away. AAPL could possibly rise above that strike price of $195 by then. If AAPL is at, say, $210 on March 22nd then we can buy it for only $195. Great, that means we would get a 'payoff' of $15; we pay $195 for the stock, immediately sell it for $210, making $15. On the other hand, if the stock is still at $170 at expiration, in fact at any price less than $195, the option will be worthless. That's because you wouldn't pay $195 for something worth less than $195. I hope.

Because of the possibility that the stock could rise enough to give a positive payoff, and because there's no downside once you own the option, then there must be some value to the option *now*.

More jargon. If AAPL is above the strike of $195 at expiry then we say that the option 'expired in the money.' If below the strike, we say it 'expired out of the money.' 'At the money' means that the stock price and the strike are the same.

Our goal now is to figure out how to value these options before expiration, when the future is unknown. That's going to involve a cunning Nobel-Prize-winning idea and some elegant mathematics. And this will result in some very counterintuitive, and interesting, results.

But, hang on, before all that, why do options even exist? One reason is that they offer a simple way to get leverage. Suppose you think that some stock, currently with value $97 is going to rise to $105 over the next month. We could buy the stock, and if we are right we'd get a return of 8.25%. Not bad for one month. There is the downside, however, that if we are wrong and the stock falls to, $92, say, that's a 5.15% fall. On the other hand, if we bought a call option expiring in one month with a strike of $100, for example, then it might cost us $1, that's the premium. If we are right, the option payoff would be $5, a phenomenal 400% return! But if wrong, … that's a 100% loss. Ouch! Are you feeling lucky?

The Mathematics

Payoff diagrams

Below is what is called a 'payoff diagram.' It shows how much a call option contract would be worth on the day of expiration as a function of the then stock price. It's very simple to explain. If the stock is out of the money, so that the share price is below the strike price, then the option is worthless. But if the share price is above the strike then you would pay the strike, receive the stock, sell the stock, and pocket the difference between the share price and the strike. That's what the plot shows. In this example, the strike is 100.

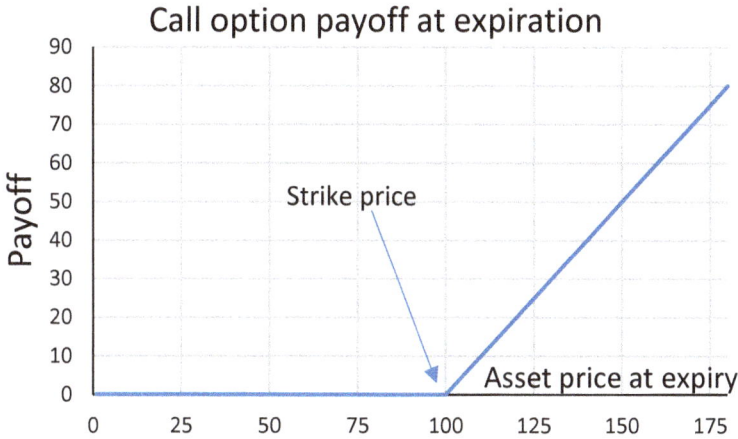

Let's use the symbol K, to mean the exercise price, T the expiration date, and S_T the value of the underlying asset at expiration. The payoff function is then

$$\max(S_T - K, 0).$$

If the stock price at a time before expiration, at time t, is S then what we want to do is figure out how much is the option worth as S and t change. In non-math terms, the option expires in the future, what is it worth today?

We can make a few simple observations. For example, the larger S is then the more the option is worth. Why? Because the higher the stock price is today, the more it will probably be worth at expiration, and thus the larger the payoff.

If the underlying is really small then there's only a tiny, tiny chance that the stock will exceed the strike at expiration, so we would expect the option value to be zero when the underlying is zero.

What else? The longer the time to expiration, that's $T - t$, the more the option is perhaps worth. Maybe? Why? Because there's more time for the stock to rise above the strike. If we are close-ish to the strike, but below it, and expiration is tomorrow then there's simply not enough time for the stock to rise. But if expiration is a year away there's plenty of time for the stock to rise.

However, on the other hand, the further away is expiration then the less the payoff is worth in present-value terms, recall the time value of money. Ok, so time to expiration is important but it's not clear quite how it affects the value of an option.

Anything else? The value can't be too high. The option to buy can't be more valuable than the stock itself.

Such observations tell us a little bit about the value of the call option, but we are still a long, long way from pinning it down to an actual number.

It's time to bring out a stock-price model.

The binomial model again

We are going to work with the same binomial model that I introduced earlier, but first with some numbers instead of symbols. And I'm going to assume that we are one day away from expiration of a call option, that the share price is exactly 'at the money' and that tomorrow, expiration, the stock could be above or below the strike.

In numbers: the strike is 100; the stock price is 100 today; tomorrow it could rise to 101 with probability 0.6 or fall to 99 with probability 0.4. I have deliberately chosen to have a biased random walk, for reasons that will become apparent. And we'll have a zero interest rate so there's no complication due to present valuing money. Here's the picture.

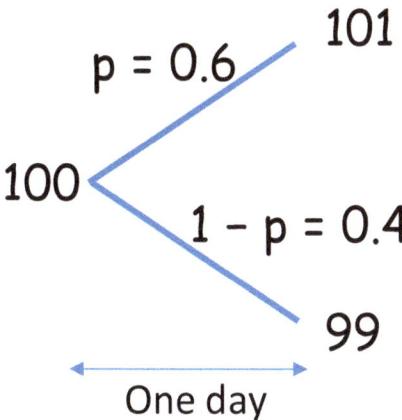

Because the random walk is biased, the expected stock price tomorrow is 0.6 x 101 + 0.4 x 99 = 100.2, so an expected increase. (If you fancy calculating the expected return for this stock when compounded over one year you'd see that this return is extremely good!)

In the next plot I've added, in red, the payoff for the call option. If the stock is at 101 at expiration then its payoff is one. If the share price is 99 then the call option expires out of the money and the payoff is zero. We don't need a model for the option value at expiration, it's obviously just the payoff. I've also added the V, with a question mark. That's what we are trying to figure out. How much is the option worth today, just before expiration?

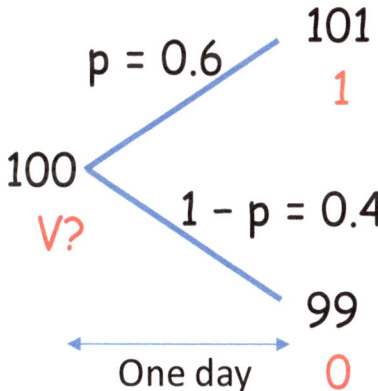

Do we have enough information to value the option today, to find V?

Normally at this point, if I were lecturing, I would ask my audience to tell me what they expect the value V should be. I would say, and I quote, 'The stock has a probability of 0.6 of rising and giving a payoff of 1, and a probability of 0.4 of a payoff of zero. That's a probability of 0.6 of getting one, otherwise nothing. What do you think the option value might be?'

Quite frankly, I must confess to you, dear reader, that I am deliberately trying to mislead my audience into giving the answer 0.6. On the grounds that that is the *expected* value, 0.6 x 1 + 0.4 x 0 = 0.6.

And that's just so I can say …

HA! WRONG!
It was a trick question!

My apologies for telling you off in all capitals, size 16 font. I realise that such behaviour is not standard teaching practice, but there are subtleties coming up and I wanted to shock you into readiness!

The reason it is wrong is that taking simple expectations is not necessarily how to *value* financial contracts. Earlier I said that people are usually risk averse, meaning that if something is risky, as is the option, then they would expect a higher return than risk free (here zero anyway). If an option's value and its expected payoff were the same then there'd be no incentive to buy the option.

No, 0.6 is not the *value* of the option, you'll be seeing the valuation done properly soon. But, yes, it is the expected payoff for the option. Tuck that idea away somewhere safe: you've got an expected payoff and you've got a value, they typically aren't going to be the same. The same is true for the stock. The stock has a value today of 100, but an expected value tomorrow of 100.2. That didn't raise any eyebrows, so it also shouldn't when I say that taking a simple expectation is not how valuation works.

How you really value options

The clever, and Nobel Prize winning, argument for valuing options is via a specially constructed portfolio consisting of one call option, and short some stock. This is how it goes.

Imagine you hold one call option, with as yet unknown value V. And you have sold short half of one share. (Technically you can't buy or sell half a share, but since everything is linear, and you could buy two options and sell one share, the following argument still works.) We get the following picture for this portfolio.

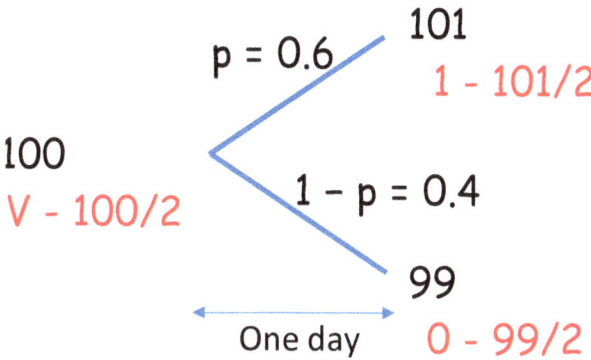

Everything in red means the portfolio. So today, on the left, the portfolio value is the option value **V** minus (we've sold) one half of the stock (which has a value of 100 now), hence **V − 100/2**. On the right, that's tomorrow, the expiry, we have: if the stock rises, the option payoff of 1, and minus one half of the stock (which now has a value of 101), that's the **1 − 101/2**; if the stock falls, the option payoff of zero, and minus one half of the stock (which now has a value of 99), that's the **0 − 99/2**.

'So?' I hear you say. Look at that picture again, below, after I've done the subtraction

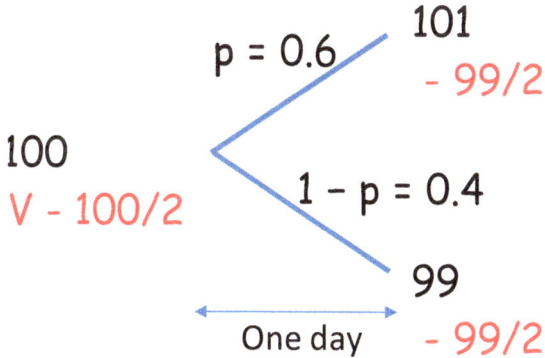

What do you notice about the portfolio values at expiration, in red on the right? They are the same. Whether the share price goes up to 101 or down to 99, the portfolio value at expiration is -99/2 in both scenarios.

What we have done here is to construct a portfolio that is risk free.

The stock is risky? Check. You could make or lose money.

The option is risky? Check. Again, you could make or lose money.

But a cleverly chosen portfolio, long one option and short some stock? Not risky. That's because the share and the option on the share are perfectly correlated, we can exactly offset the risk in one with the risk in the other.

You are probably wondering how I knew that selling half of the share would make this work, and why wasn't it two thirds, say. My answer to that is, I've been using this example for decades and it has always worked! But seriously, I'll show where that number comes from in a moment.

But once we've got a portfolio that is risk free at some time in the future it is easy to value it today. It's no different from putting money in the bank or borrowing it. Tomorrow we owe 99/2 (that's what the minus sign means), so today we also owe 99/2. (The numbers will need tweaking if interest rates are non zero, to allow for the time value of money.) This is the no-arbitrage argument. Remember, there's no such thing as a free lunch.

This means that the portfolio value *today* is -99/2, and so

$$V - \frac{100}{2} = -\frac{99}{2}.$$

On rearranging, we get

$$V = 0.5.$$

And that is our option value!

It's not 0.6. It didn't involve taking expectations. In fact, and this is very counterintuitive, we didn't use the probabilities of 0.6 and 0.4 at all in our calculations! So it doesn't matter what the probabilities are for the stock to rise or fall, we always get the same answer.

This idea of hedging was first published by Fischer Black, Myron Scholes and Robert C. Merton, but using a mathematical technique called stochastic calculus. The binomial version here is due to John Cox, Steve Ross and Mark Rubinstein.

Hedging and no-arbitrage

The are two important aspects to this valuation method/trick.

The first is hedging. I've mentioned hedging before. It's reducing risk by exploiting correlations. If correlations between two investments are not exactly plus or minus one then hedging is imperfect. Yes, you might be able to reduce risk but you can never get rid of it completely. However when you have perfect correlation, either plus or minus one, if you are clever you can eliminate risk entirely.

But why are the share and its option perfectly correlated? That's because the only random quantity affecting the value of the option is the share price. There's no other source of randomness, so they must be perfectly correlated. To take a silly example, if the option payoff also depended on the amount of rainfall at expiration then that's another random quantity affecting the option's value, and it's not a quantity that you can hedge!

The second important aspect to the method is the idea of no arbitrage. Again, I've mentioned this earlier. Once we've set up our clever risk-free portfolio, one option and short half of a share, we know exactly what we will be getting at expiration. But if we know the value of what we are getting it doesn't matter what form that value takes. It could be this special portfolio or it could be simple cash. Same value, both with no risk, there's no real difference in money terms. So they must both be worth the same today. If they had different values today then we'd just buy the cheaper one, sell the more expensive, and make a guaranteed, risk-free, profit.

That guaranteed profit is an example of arbitrage. But such situations don't exist for very long, before someone spots what's going on, takes advantage of it, and via the effects of supply and demand, cause the arbitrage to disappear. In other words, the option value would move rapidly so that the arbitrage disappears.

The idea of hedging isn't only a 'trick' as I've called it. In fact, it's not a trick at all, although it sometimes feels like it. Traders hedge options all the time.

For example, if a trader does spot a mispriced option then they would buy or sell it, depending on whether they thought it too cheap or too expensive, and immediately hedge it as above to lock in a profit. Other people in banking invent new financial contracts, sometimes called exotic options, sometimes created just for one client. They invent new payoffs, then they value the contract, then they add a bit for their profit. And they sell it to the client. The client is taking a financial risk in buying this contract, as is the seller. But the seller doesn't want the risk, they just want the profit they built in. So they hedge this exotic option, thus eliminating the risk, and locking in that profit.

The reality is a bit more complicated than all this. Isn't it always? One obvious reason is that we don't really know the model for the underlying asset! We won't worry about that in this book, that's very much an advanced topic.

The model in symbols

Now it's time to go over to symbols. This will show you how I knew how to hedge using short one half of the share. And you'll also see something unusual happening.

In all symbols we have the following picture.

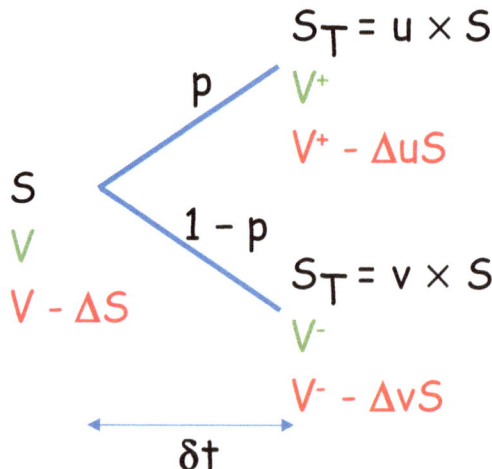

This is a very busy picture. But at least it is colour coordinated.

Anything in black is the model for the underlying. The share price today is S. (Before it was 100.) The timestep is δt. (Previously it was one day.) At expiration the asset price could go up to u times S, we just write uS, or down to vS. (Previously 101 and 99, so that $u = 1.01$ and $v = 0.99$.) The probability of going up is p, and down thus $1 - p$. (Before it was 0.6 and 0.4.)

(Note that δt is one symbol, it's not δ times t. It's said, 'delta t,' and is a very common symbol for a very small timestep.)

Anything in green is the option. The option value today is V. That's what we are trying to find. At expiration the payoff is V^+ if the underlying rises, and V^- if it falls. (Previously 1 and 0.)

And anything in red refers to the special portfolio. The only new symbol here is Δ, the Greek capital D, called Delta. (Before this was 0.5. I told you what it was, without explanation. But now I am going to show you where it comes from.)

BTW don't be put off by the use of V and v to mean completely different things!

Hedging in symbols

Hedging in this model means choosing Δ so that the portfolio value is the same regardless of whether the underlying rises or falls.

We want Δ such that

$$V^+ - \Delta u S = V^- - \Delta v S.$$

On rearranging we get

$$\Delta = \frac{V^+ - V^-}{(u-v)S}. \qquad (2)$$

This is just the ratio of the difference in the option payoffs to the difference in the asset prices. For our numerical example, this would be $(1 - 0)/(101 - 99) = 0.5$. As promised, that's where that came from.

With this choice for Δ, the portfolio value is going to be

$$\frac{uV^- - vV^+}{u - v}$$

at expiration, whether the asset goes up or down. (With our previous numbers this becomes $(1.01 \times 0 - 0.99 \times 1)/(1.01 - 0.99) = -99/2$.)

And thanks to the idea of no-arbitrage, we can then equate this value to the portfolio today to give

$$V - \Delta S = V - \frac{V^+ - V^-}{(u-v)S}S = \frac{uV^- - vV^+}{u-v}.$$

Upon rearranging, we find that

$$V = \frac{1-v}{u-v}V^+ + \frac{u-1}{u-v}V^-. \tag{3}$$

(You can check that for our numbers example this works out at 0.5.)

There are lots of things to say about this Eqn (3)!

Lots of things

The first observation is that the coefficients of the two payoffs add up to one.

$$\frac{1-v}{u-v} + \frac{u-1}{u-v} = 1.$$

The second observation is that if we compare Eqns (2) and (3) for the Delta and the option value respectively, they are completely different formulae. So when we previously found 0.5 for both the hedge ratio (the Delta) and the option value that was a coincidence. You might have thought that because we hedged with 0.5 of the underlying asset that that was why we got an option value of 0.5. No. Pure coincidence. That's the beauty of using symbols, it shows us structure so much better than numerical examples do.

The third observation is as before, there's no **p** in the results, neither for the Delta, nor for the option value. We really don't care whether the stock rises or fall, thanks to hedging. We do, however, care where the asset rises or falls *to*, that's in the **u** and the **v**.

For the fourth observation I'm going to introduce a new symbol, **p'**:

$$p' = \frac{1-v}{u-v}.$$

In terms of this new symbol, we can write Eqn (3) as

$$V = p'V^+ + (1-p')V^-. \tag{4}$$

I've used the symbol **p'** because it rather looks like a probability. Indeed, in Eqn (4) the two numbers in front of the payoffs do add up to one as discrete probabilities should. So we can think of **p'** as a probability, in which case today's option value *is* an expectation of future option values. This is precisely what I earlier said it wasn't! I said you can't take simple expectations to value options, because of all that risk-aversion stuff. That is true. You can't take *simple* expectations. But there is a certain special probability that you can use to value options. It's just not the real probability. Just to be absolutely clear on this I'm going to call it a *wrong* probability.

This is now where the turbulence starts, and why you need those seatbelts.

Risk-neutral probabilities

Don't worry if you get confused by what follows. In my experience most people who do university degrees in mathematical finance struggle with this particular topic.

Step away from options for a short while. Suppose I said imagine you are in a world in which they don't care about risk. A world in which everything is valued by taking expectations. A world in which no one needs to be financially compensated for taking risk. We might call that a risk-neutral world, because risk neutral means precisely all that, that we don't take risk into consideration when we make financial decisions or value financial instruments.

So we are in that risk-neutral world. That doesn't mean that we don't know about randomness or probabilities though. We could ask a simple question, 'What must the probabilities be in our random walk so that today's share price is valued by an expectation?'

Take our earlier numerical example, in which the stock is 100 today, but could rise to 101 or fall to 99. What must the probabilities of the rise and fall be in order for that 100 to be the expected value for the future 101 and 99? I think it's fairly obvious, thanks to symmetry, that the 101 and 99 must be equally likely. That's because 0.5 x 101 + (1 - 0.5) x 99 = 100. We call that 0.5 in this case the 'risk-neutral probability.' (It won't always be 0.5, of course. It depends on the up and down moves.)

Notice that's it's very different from the 0.6 and 0.4 that we presumably got from some complicated statistical analysis of past asset data. Those probabilities are the *real* probabilities. And the *real* expected stock price tomorrow was 100.2.

Risk-neutral probabilities don't exist. But they answer that interesting question about what they should be in that risk-neutral world.

Let's do that same calculation using the symbols. Remember that the asset today is S, after the timestep it is either uS or vS. And let's call the risk-neutral probability of the stock rising p'' (we say 'p double dash').

The calculation we have to do is the same as in (5) but with the stock instead of the option:

$$S = p''uS + (1 - p'')vS.$$

Dividing throughout by S and rearranging and we get

$$p'' = \frac{1 - v}{u - v} = p'.$$

Pardon my French, but OMG! This risk-neutral probability is the same as the 'wrong probability' we used in our correct-thanks-to-delta-hedging derivation of the option value.

What can this possibly mean?

Let me go through what this means and how it is used. Slowly.

Imagine, if you will, that you inhabit a world in which no one cares about risk. They do care about what happens on average, but not at all about deviations about that average. To them every roll of the die might as well be 3.5. Do they care that rolling a one means they don't move very far in the game of Snakes and Ladders? No. Do they care that a three will land them at the bottom of a ladder? No. And that a five will land them at the top of a very long snake? Equally no.

In the world of stocks they know that the stock could go up to 101 or down to 99, or, in symbols, up to uS or down to vS. But the probabilities? They don't care. Not only do they not care, they can't even be bothered to look at past stock prices to figure out statistically what those probabilities might be.

They assume, incorrectly as far as we, more sane, risk-averse people are concerned, that therefore the current stock price must be the expected future stock price.

Now they know what the stock is now, and what it could be in the future, and given their (I'll say it again, wrong) assumption, this means that they have enough information to calculate some probabilities. That's exactly what I did to calculate the risk-neutral probability of 0.5 with my numbered example or $(1 - v)/(u - v)$ in symbols.

Once they have found this probability, the *wrong* probability, they then use this to calculate the value of options today using Eqn (4).

And even though they've used the wrong probabilities, *they still get the right answer*!

This is about the only time in your life when two wrongs do make a right!

I sometimes explain this by saying that one wrong reverses the other wrong.

In the exercises I will ask you to work out some risk-neutral probabilities, to help you get the hang of all this.

Non-zero interest rate

If the interest rate is not zero then there are only a couple of tiny changes to be made to the above hedging and risk-neutrality arguments. I won't go into the details, I don't think they add much to the understanding of the concepts, but there is a danger that they will be a confusing distraction. However, in words, all you have to do is remember to take the present value of any payments in the future. It doesn't affect the hedging, but it does mean that you must discount the risk-free portfolio payoff. And when calculating risk-neutral probabilities you also need to discount the 'expected' share. That means that the risk-neutral probabilities will change a bit.

The role of the asset's growth rate and volatility

We've seen that the real probabilities of the share pricing rising or falling have no effect on the value of an option. This has another consequence that I will just state. And that is that the value of an option does not depend on the expected growth of the stock, μ. Yes, its value depends, very importantly, on the volatility of the stock, σ. But thanks to hedging, we don't care whether the stock rises or falls, and so its average move is irrelevant!

Trees

I've given a numerical example in which I said we were 'one day away from expiration.' Then with the symbols I introduced a timestep. But in both cases the future was represented by just two possible share prices. Now that's not very realistic. It's great for explaining how valuation and hedging are done and the concept of risk neutrality, but when valuing options you need something a bit more representative of future share prices. However, that's easy with the binomial model. All we have to do is stack each binomial one after another. And this is shown in the next picture. This is called a binomial tree.

In this plot, the share price is currently 100. It can go up to 1.01 x 100, or down to 100/1.01. That's one timestep. After the next timestep the share price can be at 1.01^2 x 100, 1.01 x 100/1.01 = 100 or $100/1.01^2$. And so on, for more timesteps, right up to expiration. Notice how I've used slightly different numbers from before. The down move is by a factor of 1/1.01 this time, not the previous 0.99. That's simply so I get a nice tree that always has one possible asset value of 100 after even numbers of timesteps. It's not crucial to do this though.

I've shown some of the formulae in this diagram. In each case they are the formulae quoted above for the Delta and for the option value, the latter in terms of the risk-neutral probabilities.

I've also colour coded the cells. Yellow means it's an input I can change. The pinkish cells are the share prices as we move along the tree. Pale green are the option values. And pale blue are the Deltas. The last column represents expiration so the option formula is just the payoff. I haven't put in the formulae for every cell, they are repeated throughout the tree, only expiration is different.

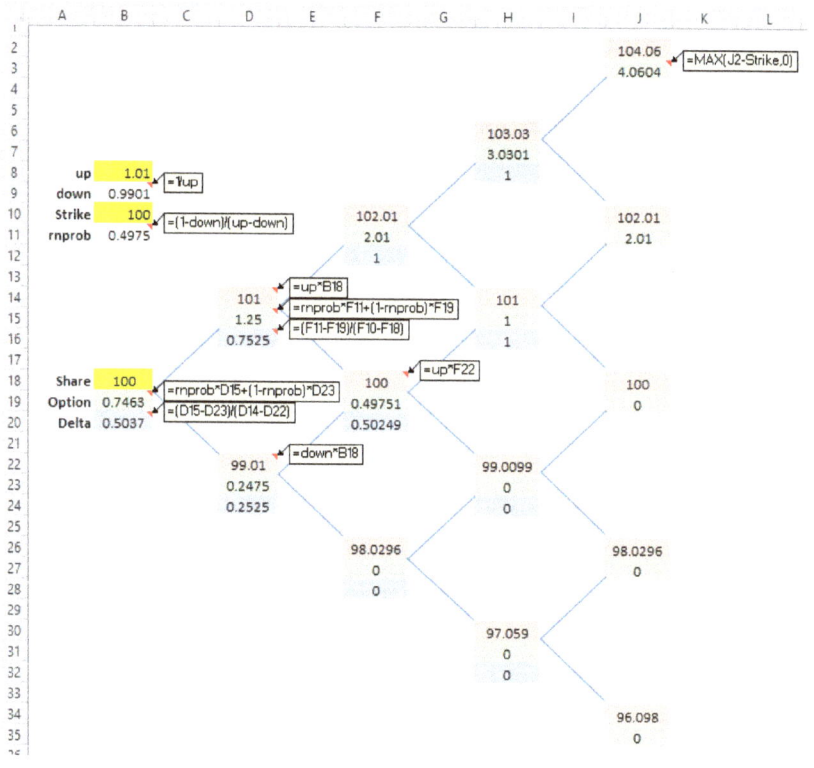

Put options and other payoffs

Call options, the option to buy a share, aren't the only type of option. There are many other payoffs possible. The other common one is the put option, the right to *sell* the share in the future.

The payoff diagram for the put option is below. It's easy to see where this comes from. When we get to expiration we have to make a decision whether or not to exercise the option. If the share is above the strike at expiration then it wouldn't make sense to exercise the option. You wouldn't sell the asset for K if you could sell it for S > K. So the payoff for S > K is zero. On the other hand, if S < K then we would be better off selling it for K, rather than S, and that means exercising the option. We would buy the share for S, sell it immediately for K, making a profit of K - S > 0.

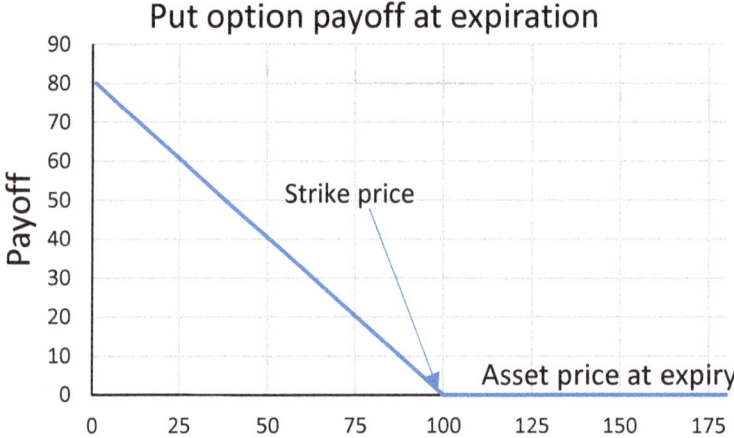

The payoff for this option is

$$\max(K - S_T, 0).$$

If call options are a possibly good investment choice if you believe a stock is going to rise and you want some simple leverage, then put options might be a good investment if you believe a stock is going to fall. You could also profit from a falling share price by selling a stock short. But that suffers from the obvious problem that if you are wrong and instead the share rises there's potentially no limit to how much you could lose. With a put option your maximum loss is whatever you paid for the option in the first place, its premium.

Another reason for buying put options is for insurance. You might not believe that the shares you currently own are going to fall in value but you perhaps can't afford to take that risk. Rather than selling your shares and missing out on any growth, you buy put options that will pay off if they do fall. They will give you insurance up until they expire.

You can value put options in a similar way to call options, by setting up a special hedged portfolio. Eqns (2), (3) and (4) are all still valid, you just need to put in different values for the payoff at expiration.

But there is also a simple relationship between the values of call and put options, so that if you know the value of one you can easily find the value of the other.

Put-call parity

I'm going to show you another example of the no-arbitrage principle and in so doing show you how to value a put option based on the value of a call option with the same strike. And this involves setting up another portfolio.

Imagine buying one call option with a strike of K. And simultaneously selling a put option with the same strike.

The payoff for this option portfolio is thus

$$\underbrace{\max(S_T - K, 0)}_{\text{For the call}} - \underbrace{\max(K - S_T, 0)}_{\text{For the put}} = \underbrace{S_T}_{\text{Stock}} - \underbrace{K}_{\text{Cash}}.$$

Why? Because if $S_T > K$ then the first term is $S_T - K$ and the second is zero. But if $S_T < K$ then the first term is now zero and the second is $-(K - S_T)$ which is again $S_T - K$.

The left-hand side is the payoff for a long call and the short put. The right-hand side is the stock minus cash. We can rearrange this

$$\underbrace{\max(S_T - K, 0)}_{\text{For the call}} - \underbrace{S_T}_{\text{Stock}} + \underbrace{K}_{\text{Cash}} = \underbrace{\max(K - S_T, 0)}_{\text{For the put}}.$$

And this means that a put option has the same payoff as long one call, short stock and some cash. And again, thanks to the idea of no arbitrage, this relationship must be true at all times before expiration.

$$\text{Put} = \text{Call} - \text{Stock} + \text{Strike in Cash}.$$

Ta da! The value of a put option is the same as that of a call with the same strike and expiration minus the current stock price plus the strike price in cash.

Expectations, a recap

We've seen expectations many times. Sometimes I've told you that we take expectations to calculate something useful like an expected stock price, given the real probabilities. But sometimes I've warned you that you mustn't take expectations, that can lead to the wrong answer if you use the wrong probabilities. Well, we are going to recap all that here.

Stock price 100, tomorrow goes up to 101 with probability 0.6, or down to 99 with probability 0.4. A call option payoff will be one or zero. Let's look at some expectation calculations for the stock and the option.

Case 1: Real expectation for the stock, real expectation for the option
Real expected stock price tomorrow: 0.6 x 101 + 0.4 x 99 = 100.2. Correct!

Real expected option payoff: 0.6 x 1 + 0.4 x 0 = 0.6. Correct! But it's *not* the value of the option today!

Case 2: Risk-neutral expectation for the stock, risk-neutral expectation for the option
This is all about option *valuation*.

Throw away those real probabilities, the 0.6 and the 0.4!

What must the probabilities for the stock price up and down moves be if today's price is to be the expected price tomorrow?

Since 0.5 x 101 + 0.5 x 99 = 100, the probabilities must both be 0.5. These are the correct *risk-neutral* probabilities! But they aren't real.

The option value is then 0.5 x 1 + 0.5 x 0 = 0.5. Correct! This is the option's theoretical value, it's *not* its expected payoff though!

What Next?

Again, expect pretty much everything to be in continuous time in university. Instead of using the binomial model and binomial trees to value options, you will use stochastic calculus. There will still be the same hedging process but in continuous time this leads to a famous partial differential equation, the Black-Scholes equation. Solve that equation and you'll arrive at the Black-Scholes option pricing formulae.

You will get to value options that are more complicated than simple calls and puts.

You might also get to address one of the big problems with simple option theory, and that is how share price volatility, the parameter crucial to option valuation, is not a nice constant. Part of that is the subject of stochastic volatility. Think rolling a die, but with every roll the number of spots, indeed the number of sides, changes. Mind. Blown.

Exercises

1. A call option has a strike of $120. At expiry the underlying asset is $115, what is the payoff? What is the payoff if the asset is $120, or $135?
2. A share price is $87. If the share price falls today to $86 would a call option with two months to expiry go up or down in value?
3. Which is worth more, a call option with a strike of $75 or a call option with a strike of $70? Both options have the same underlying and the same expiration.
4. A stock is currently $56. Tomorrow it could rise to $58 with real probability 75%, or fall to $55 with probability 25%. What is the *real* expected share price tomorrow?
5. A share price is $80. It could rise to $82 or fall to $77. It has a *real* expected price of $81. What is the probability of it rising?
6. Tomorrow a share could be worth $53 with probability 55% or $49 with probability 45%. What is its value today? (This is a trick question!)
7. Tomorrow a share could be worth $53 with probability 55% or $49 with probability 45%. What is its real expected value? (This is not a trick question!)
8. A share has value $127.34 today. Tomorrow it could rise to $129.61 or fall to $124.99. A call option on this stock has strike $127 and expires tomorrow. What are the payoffs for the option if the stock rises, and if the stock falls? How can you hedge this option? If the risk-free interest rate is zero, what is the value of this call option today?
9. Go back to Question 8. What is the risk-neutral probability of the share rising? Use this to value the option. Is your answer the same as before?
10. Go back to Question 8. What is the hedge ratio, the delta, for a put option with the same strike and expiration? What is the put option's value? Use both the hedging and risk-neutrality methods, the answers should be the same.
11. Go back to Question 8. What is the option value if the one-day, risk-free, interest rate is 0.0397%?
12. Interest rates are zero. A stock has value $223. A call option with strike $230 and expiration next month has a value of $4. What is the value of a put option with the same strike and expiration?

APPENDIX

In this Appendix I am going to look at some of the mathematics that will help you take you to the next level on your mathematical finance journey!

Exponential Function

Two times two. Two squared. Two to the power of two. These are all

$$2^2 = 4.$$

Two by two by two. Two cubed. Two to the power of three:

$$2^3 = 8.$$

Two to the power of **n**:

$$2^n.$$

As a function of **n** this looks like this:

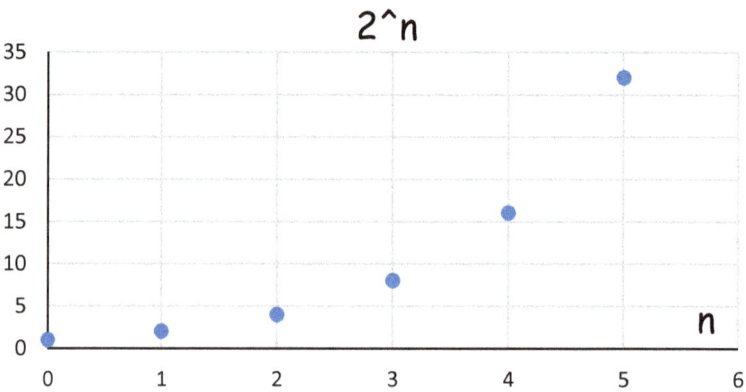

So far, so good. What about, say, 2.718 squared, cubed, etc.? Well, you know how to multiply non integers, so they're easy enough. But what about raising numbers to *non-integer powers*? Something like

$$2.718\ldots^{3.901},$$

for example. What we are now looking at is a special function called the exponential function. It's written like this

$$e^x \text{ or } \exp(x).$$

And looks like:

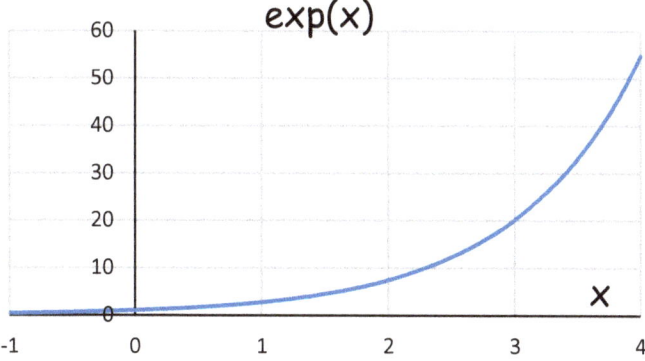

If you have a scientific calculator you will find the exponential function on it.

Properties of the exponential function

We can raise any number to any power, but when the number being raised to some power is 2.718282... = e then the function has special properties. One special property is that when you measure the slope of this function, it is also e^x.

What about raising e to the power $x + y$?

$$e^{x+y} = e^x \times e^y.$$

What was an addition now looks like a multiplication!

Another useful fact about the exponential function is that it can be written as

$$e^x = 1 + x + \frac{x^2}{2!} + \frac{x^3}{3!} + \frac{x^4}{4!} + ... = \sum_{i=0}^{\infty} \frac{x^i}{i!}.$$

That's an infinite sum! But one that converges for all values of x. Such an expansion in powers is known as a Taylor series.

Logarithms

We know how to raise numbers to powers, and I've said that raising the number **e** = 2.718282... to a power is rather special. But what if I turned that around and asked, 'What power must we raise **e** to get, say, 45.60?'

The answer requires us to know about another special function, called the natural logarithm, written

$$\ln(x).$$

The answer to the question 'What power must we raise **e** to get, say, 45.60?' is then ln(45.60), which is 3.82. That just means that exp(3.82) = 2.718...$^{3.82}$ = 45.60.

The exponential and logarithmic functions are inverses of each other. But by inverse I don't mean 'one over,' perhaps *reverses* is a better word. So, exp(3.82) = 45.60, and then ln(45.60) = 3.82. Generally,

$$\ln(\exp(x)) = x \text{ and } \exp(\ln(x)) = x.$$

The natural logarithm function looks like this:

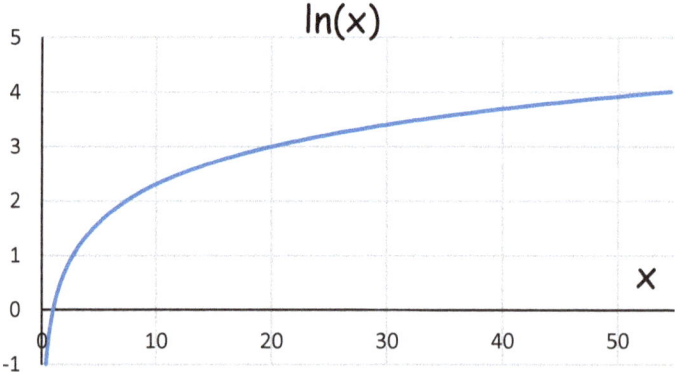

If you have a scientific calculator you will find the logarithmic function on it.

Because the exponential and logarithmic functions are inverses of each other the plots are reflections of each other about a diagonal line going through the origin.

Properties of logarithms

One other feature about logarithms that you will often need is what happens when we take the logarithm of a product, xy?

Remember that

$$x = e^{\ln(x)}.$$

And

$$y = e^{\ln(y)}.$$

So

$$xy = e^{\ln(x)} e^{\ln(y)} = e^{\ln(x) + \ln(y)}.$$

Now take logarithms of both sides and we get

$$\ln(xy) = \ln(x) + \ln(y).$$

Which turns a multiplication into a sum! In words, the logarithm of a product is the sum of the logarithms. (Now ask your grandparents about slide rules and log tables!)

Similarly, when we take the logarithm of a power this is what happens:

$$\ln(x^a) = \ln\left(\left(e^{\ln(x)}\right)^a\right) = \ln\left(e^{a\,\ln(x)}\right) = a\ln(x). \quad (A1)$$

I've greyed out the middle step to emphasise the important result.

There's also a Taylor series expansion for the logarithmic function, looking like

$$\ln(1 + x) = x - \frac{x^2}{2} + \frac{x^3}{3} - \frac{x^4}{4} + \ldots = \sum_{i=1}^{\infty} \frac{(-1)^{i+1} x^i}{i}. \quad (A2)$$

(This series only converges for $-1 < x \leq 1$.)

Continuous Compounding

Back to fixed income. Now let's look at what happens when the number of interest payments gets larger and larger while the size of those payments gets smaller and smaller. The reason for this is that if you ever study finance at a higher level you will find that most of the mathematics is in what's called 'continuous time.' Interest rates, for example, are not treated as quantities that change in value every year, say, but are constantly evolving. This gets us into calculus and differential equations, but that's a bit too heavy for this book. However, there is a little bit of continuous-time finance that we can do.

It's time to bring in those special functions that I described earlier, the exponential and logarithm.

Recalling that these functions are inverses of each other, we can write

$$\left(1 + \frac{r}{n}\right)^{nT} P = P \exp\left(\ln\left(\left(1 + \frac{r}{n}\right)^{nT}\right)\right) = P \exp\left(nT \ln\left(1 + \frac{r}{n}\right)\right).$$

This has used Eqn (A1) on Page 99.

Now I'm going to ask the question, 'What happens as the number of interest payments gets larger and larger?'

Mathematically, this means how does the above change as n gets bigger and bigger? We write this as $n \to \infty$. And we say, '… as n tends to infinity.'

As n gets larger and larger then r/n gets smaller and smaller, and so, remembering Eqn (A2) on Page 99, the Taylor series expansion of the logarithm function,

$$\left(1 + \frac{r}{n}\right)^{nT} P = P \exp\left(\ln\left(\left(1 + \frac{rT}{n}\right)^{nT}\right)\right)$$

$$= P \exp\left(nT \left(\frac{r}{n} + \cdots\right)\right).$$

Here the '…' means 'small terms that we can ignore.'

As n gets larger and larger, we get to

$$\left(1 + \frac{r}{n}\right)^{nT} P \sim P \exp(rT) = P\, e^{rT}.$$

(The ~, the tilde symbol, is in this sentence a verb. And mathematicians say 'twiddles.' Really, I kid you not! It sounds like the name for a very cute cat ... 'Here, Twiddles!' It means that the left-hand side tends to, i.e. becomes closer to, the right-hand side, in this case as n tends to infinity. Say 'Blah blah twiddles blah blah.')

Now you can see why I wanted to introduce you to the exponential function. That's because as interest payments get more and more frequent your money in the bank or in a bond grows 'exponentially.' Infinitesimal interest payments paid infinitely often is called continuous compounding. And is represented by the exponential function.

Annual Rate Versus Continuous Compounding

Suppose you get an interest rate of r in one year. What is that equivalent, call it r^*, in terms of continuous compounding? You need to do the following:

$$1 + r = e^{r^*}.$$

Or rearranging,

$$\ln(1 + r) = r^*.$$

You can see that r and r^* are not the same.

r^* (continuous)	r (annual equivalent)
1%	1.005%
2%	2.020%
3%	3.045%
4%	4.081%
5%	5.127%

Floating Interest Rates

To give you a sense of some of what happens when interest rates are floating and therefore liable to change, perhaps randomly, I want to do one example with such random interest rates.

You've got $100 in a floating-rate bank account. Your deep analysis of the economy, the declarations of the central bank, your statistical examination of historical data going back to Columbus, and the pronouncements of your friendly neighbourhood haruspex (google it!), lead you to believe that the floating interest rate will change tomorrow. It will be either 4% for two years or 6% for two years. And it's 50:50 which it will be.

Let's look at how much money you might have in one and two years' time. We'll also look at expected values, that is, an average. (And that's why I've given you the probabilities.)

	After one year	After two years
Case 1: r is 4% for two years (50% chance)	$104	$1.04^2 \times 100 = \$108.16$
Case 2: r is 6% for two years (also 50% chance)	$106	$1.06^2 \times 100 = \$112.36$
Expected value	0.5 x 104 + 0.5 x 106 = $105	0.5 x 108.16 + 0.5 x 112.36 = $110.2**6**
Case 3: Take the average rate of 5% for two years	$105	$1.05^2 \times 100 = \$110.2\mathbf{5}$

Did you see what just happened? I'm guessing not, it is very, very subtle. It's all about looking at averages when interest rates are random and what that means for how much money you have in the bank.

We are comparing rates of 4% and 6%, both equally likely. So you might think that since we are looking at expectations we might as well take the average of these, it's 5%, and use that.

After one year that works. With a rate of 4%, after one year you have $104. With a rate of 6%, after one year you gave $106. The average amount in the bank is $105. And that's the same as what you'd get using the average interest rate of 5% for the year.

Now look at what happens after two years. The average value of the two scenarios is $110.26. But if you use the average rate of 5% for two years you get $110.25. Those numbers are different.

I know, it's just one cent different. But remember the hedge fund managers are multiplying these numbers by the billions. (The bonus they thus get is the difference between a 120' superyacht and a 150' one. I exaggerate a little bit.)

A clue as to what is happening can be found in Eqn (1) on page 42. That's the formula I am using to calculate values. When $n = T = 1$ that expression is linear in r. What do I mean by linear? I mean that if we were to plot that expression as a function of r then it would be a perfectly straight line.

But when $T = 2$, when we are looking at two payments of interest compounded, it is nonlinear as a function of r. It is a quadratic function.

And what we are seeing in the one cent difference is, drum roll, ...

The average of a function is not the same as the function of the average.

(Unless that function is linear.)

Here 'the average of a function' means calculate the values in the two scenarios after two years and then average those values.

And 'the function of the average' means take the average rate and then calculate the value using it.

One method of calculating is correct (the average of a function), one is wrong (function of the average).

It's all about a clever piece of mathematics called Jensen's Inequality.

For now you just need to know that quadrillions of dollars are passing through instruments for which Jensen's Inequality is important. Get it right as a trader or quant and you will get a big bonus, get it wrong and you'll be looking for a new job.

Probability Distributions

In the examples in an earlier chapter, the numbers in my lists, or generated randomly by the roll of a die, were integers. These illustrated the idea of a *discrete* probability distribution. With discrete distributions the numbers don't have to be integers, nor do they have to come from a finite list. They don't even have to be numbers, they could be eye colours for example. Discrete probabilities can be plotted in a histogram, such as that below.

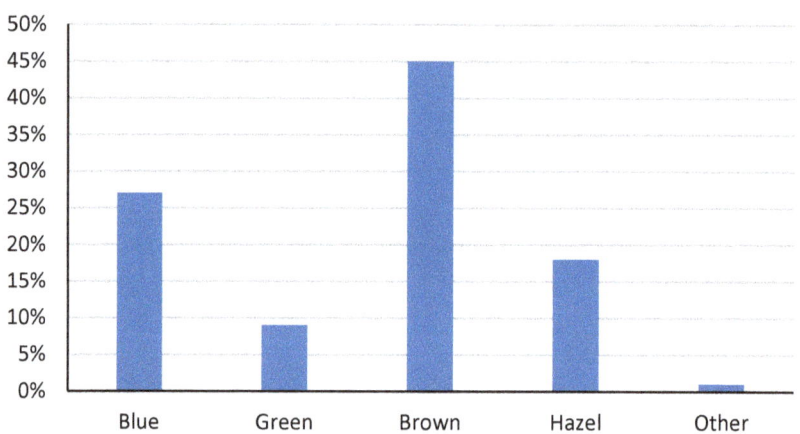

The probabilities in a discrete distribution must sum to one.

But there are also *continuous* distributions. An example would be people's heights.

Now no one is *exactly* 175cm, no matter what it says in your passport. Someone might be 174.9733… cm, the number would go on forever! And so the probability of any precise height is zero! Not very helpful. But that doesn't mean that we can't plot the distribution of heights.

Here's a plot of height distribution. This plot requires some interpretation, but it's quite straightforward.

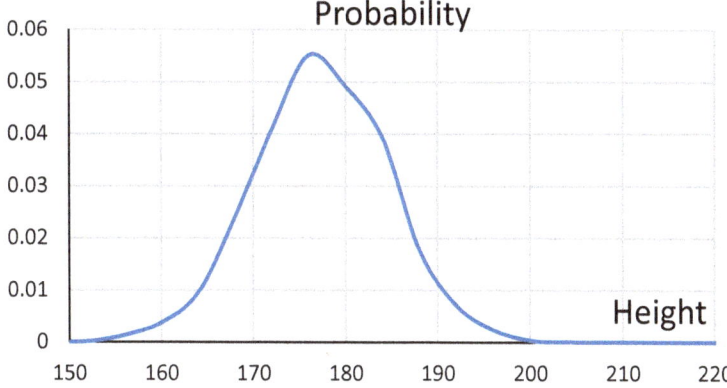

Since the probability of being exactly 175cm is zero we have to ask questions like 'What is the probability of being between 174.5 and 175.5?' This can be answered by looking at the area under the curve from 174.5 to 175.5, and it's 5.62%. See the plot below.

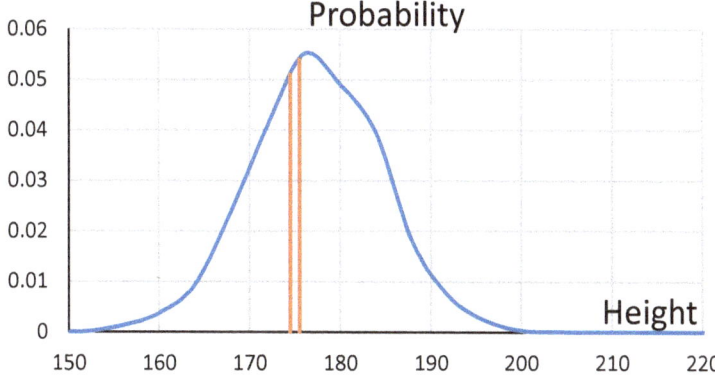

That's how we interpret the plot of continuous distributions. The function it represents is called a 'probability density function.' (The area under the curve between 175 and 175 – no, that's not a typo – is going to be zero, in agreement with what I said about no one's height being exactly 175cm.)

Now instead of the heights of the discrete distribution adding up to one, it's the area under the curve that must total one. That's simply because everyone, that's 100% of people, has some height!

The Law Of Large Numbers

There are two probability results that are useful in finance. The first is the Law of Large Numbers, and the second is the Central Limit Theorem.

Here are the results of a simple experiment I conducted rolling a die, although actually done on a spreadsheet.

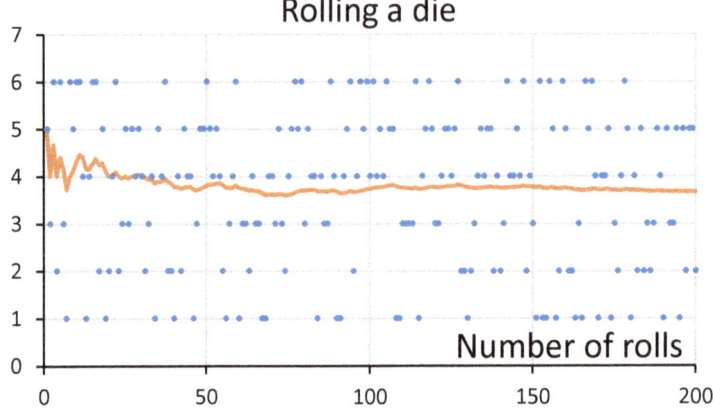

I rolled a die 200 times. The plot shows the result of each roll, the dots, against the number of rolls so far and the average, the orange line, up to that roll. (So that after, say, 37 rolls the average plotted is of those 37 rolls.)

While the dots are bouncing around randomly it is clear that the average is settling down, heading towards the average of 3.5.

And that's what the Law of Large Numbers says, that if you perform the same experiment more and more times the average you calculate will converge to the theoretical expected value. It doesn't have to be rolling a die, it could be tossing a coin, or spinning a roulette wheel, measuring people's heights or weights, amount of rainfall, etc. There are some technical conditions for this theorem to work, such as the numbers have to be independent of each other and coming from the same probability distribution each time.

Next a very important theorem in probability.

The Central Limit Theorem

We are going to do another experiment. But this time I'm tossing a coin. If I toss tails, it counts as -1, heads is +1.

But before I start, I want to remind you about Pascal's Triangle. This is the triangular pattern of numbers starting with the number 1 at the top, below this being two more 1s, then each row is calculated by adding up just two numbers in the row above. You'll recognize this picture:

```
                        1                                   1
                    1       1                               2
                1       2       1                           4
            1       3       3       1                       8
        1       4       6       4       1                   16
    1       5      10      10       5       1               32
  1     6      15      20      15       6       1           64
1     7      21      35      35      21       7       1     128
```

The column to the right is the sum of the numbers in each row, you'll see they are powers of 2.

Now the experiment.

This is what I do. I start by tossing the coin once, and writing down in a spreadsheet the result. Then I toss again, and write down the result. This is what I get.

H	1
T	-1
H	1
H	1
H	1
H	1
T	-1
H	1

I do this many, many times. I can then plot the probability of getting heads (+1) and tails (-1). Since this is a very simple experiment, I don't need to actually do all that tossing. I can go straight to the theoretical result, which is obviously 50:50, and would look like this:

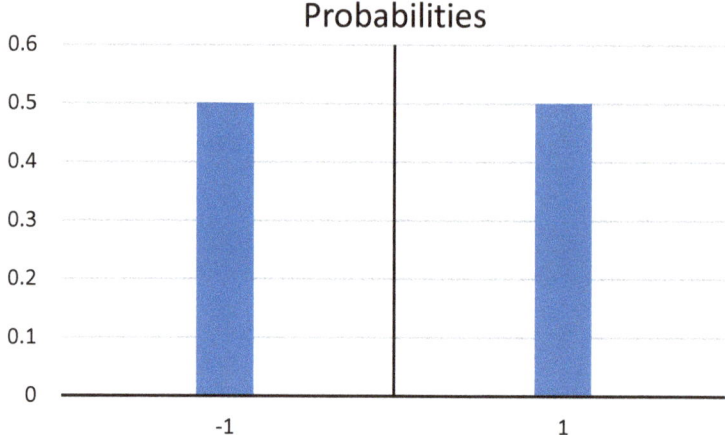

This distribution has a mean of zero and a standard deviation of one.

Next, same idea but toss the coin four times, and add up the result. Repeat. This might look like the following.

				Count
H	T	H	H	2
T	T	T	H	-2
H	H	T	H	2
T	H	T	H	0
H	H	H	T	2
H	T	T	H	0
H	T	H	H	2
T	T	T	T	-4

And then plot the distribution of the count. Again, I can do this theoretically, I don't really need to toss the coin. The theoretical answers come from Pascal's triangle. Read across the fourth row of the triangle, that's 1, 4, 6, 4 and 1, and divide each of those numbers by the 16 on the right. And those are the probabilities we need.

The Mathematics of Finance for High Schoolers

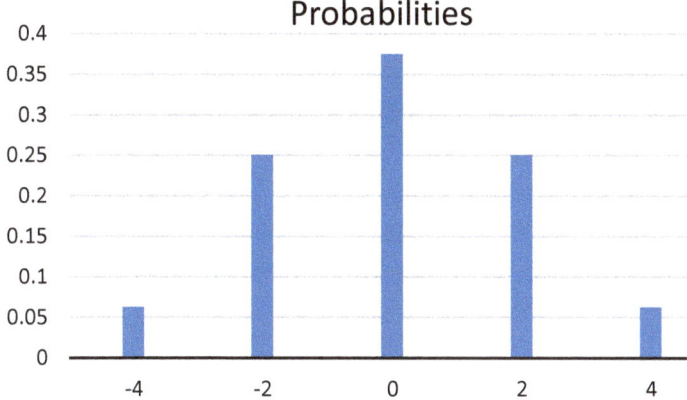

This distribution has a mean of zero, that's fairly obvious, and a standard deviation of two. (Recall that we can add variances, for four tosses that's four, giving a standard deviation of two.)

What we've done here is plot the distribution of the *sum* of four random numbers. Let me say that again. We've taken four random numbers and added them up. This gives us a *new* random number. It ranges from -4 to +4. We are now going to repeat this experiment, but this time with a few minor changes.

First, we are going to toss the coin a whopping 100 times! And add up those 100 numbers, all those -1s and +1s. This gives us one new random number, it'll be somewhere between -100 and +100. It will have a mean of zero and a standard deviation of 10. (Why is the standard deviation ten?)

Repeat this, toss the coin another 100 times and sum, this gives us another random number. Repeat, another random number. And repeat. Let's say we do this 2,639 times. So we have 2,639 random numbers, all in the range -100 to 100.

Now the second change, we subtract off the mean from each one of those 2,639 numbers. That was easy, since the mean is zero! And then divide all of those new numbers by the standard deviation, ten. This leaves us with 2,639 random numbers, a list of numbers that now has a mean of zero and a standard deviation of one.

Before I plot the distribution of these 2,639 random numbers I want to explain why I chose the number 2,639. It's always tempting to use nice round numbers in examples, but then there's the risk that you, the reader, might think there's some relationship between the numbers I chose. If I'd said generate *1,000* random numbers you might think that was related to the adding up of the results of the *100* coin tosses, and the standard deviation of *ten*. And that's not the case. The 100 and the 2,639 have absolutely nothing to do with each other. I just needed a large number, and chose 2,639. More in a moment.

Here's the discrete probability distribution for the sum of the 100 -1s and +1s, with the mean subtracted and divided by the standard deviation of ten. Again, I didn't actually toss the coin 2639 x 100 times, I just read the numbers off Pascal's triangle and divided by 2^{100}.

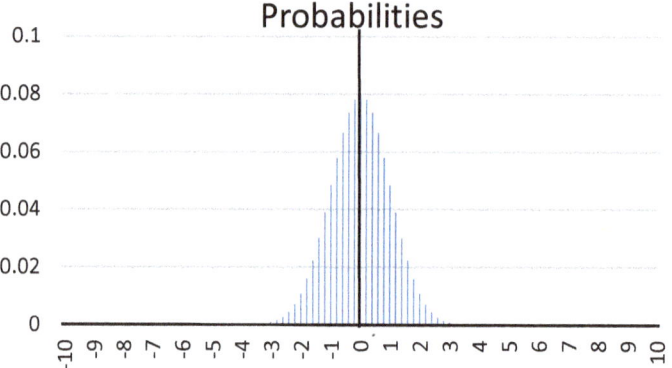

The shape of this curve is very special in probability theory. It looks like the 'probability density function' for what is called the standard normal distribution.

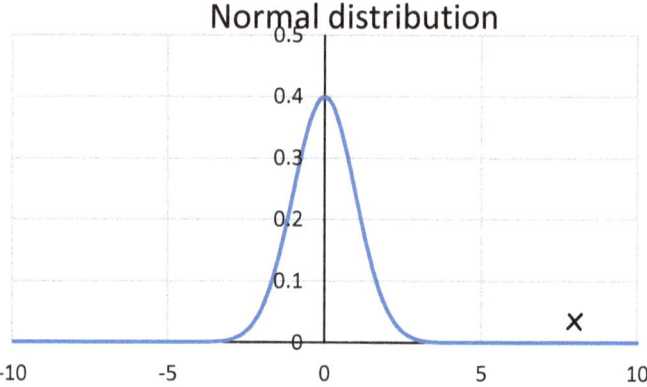

(The *shape* is the same, but the actual *numbers* are different from those in the coin-tossing experiment because the coin tossing gives a *discrete* distribution while the normal distribution is *continuous*. To get from the *histogram* whose values *add up* to one to a *curve* with an *area* of one you need to scale with the width of the bars in the histogram which is 0.2.)

The *standard* normal distribution is a continuous probability distribution, it is symmetrical about zero, has a standard deviation of one, and in the left- and right-hand tails falls very rapidly but never reaches zero. Being a continuous probability distribution the area under its curve is one.

The probability density function for the standard normal distribution is

$$\frac{1}{\sqrt{2\pi}} \exp\left(-\frac{1}{2}x^2\right).$$

That's what I plotted above.

And now after those examples I can reveal the very important Central Limit Theorem!

Let $x_1, x_2, \ldots, x_n$ be n independent random numbers drawn from a distribution having a mean of m and a standard deviation of s then as n gets larger and larger, tending to infinity, the distribution of the scaled average

$$\frac{\left(\frac{1}{n}\sum_{i=1}^{n} x_i\right) - m}{s}$$

tends to the standard normal distribution.

In English, that means that if you take a lot of random numbers from the same distribution, calculate their average, subtract off the mean of the distribution and divide by the standard deviation, then the result is close to being normally distributed.

That's a lot of 'if's. Let me say it again, a bit neater this time. But be aware that every time I try to make it sound simpler I am missing out some key technical elements. Here goes:

The average of a lot of random numbers, and it doesn't matter what sort of random numbers they are, looks like the normal distribution.

And here are the key bits missing from that simple description.

- 'A lot' technically means 'as the number of random numbers tends to infinity.'
- 'Random numbers': The random numbers have to be independent of each other.
- 'It doesn't matter what sort of random numbers': The underlying distribution doesn't matter as long as its mean and standard deviation are finite.
- 'Average': If you take off its mean and divide by the standard deviation of the underlying distribution then the resulting normal distribution has a mean of zero and a standard deviation of one.

The importance of the Central Limit Theorem is that it explains why so many quantities seem to be normally distributed. In financial markets the normal distribution is often used as a model for share prices. Perhaps adding many, many trades during a day results in share prices changes that look normally distributed.

The Normal Distribution

The *general* form of the normal distribution is

$$\frac{1}{s\sqrt{2\pi}} \exp\left(-\frac{1}{2s^2}(x-m)^2\right).$$

The general form has a mean of m and a standard deviation of s. The plot looks the same as the *standard* normal distribution, having a mean of zero and a standard deviation of one, except that is shifted left ($m < 0$) or right ($m > 0$), and squashed ($s < 1$) or stretched ($s > 1$). It has just the two parameters. That's a nice quantity of parameters to have, not too simple and not too complicated.

The normal distribution is probably the most important continuous distribution. It has several interesting and important properties.

First, it's symmetrical. And its mean, mode and median values are all the same.

Second, if you add up normally distributed random variables then what you get is also normally distributed. They don't even have to have the same mean and standard deviation, they just have to be independent. (Two random numbers are 'independent' if one random number does not affect the probability of occurrence of the other.) Suppose that we have n normally distributed random numbers with means and standard deviations m_i and s_i for $i = 1$ to n, then the mean, m, and standard deviation, s, of the sum are given by

$$m = \sum_{i=1}^{n} m_i$$

and

$$s = \sqrt{\sum_{i=1}^{n} s_i^2}.$$

Normally Distributed Share Price Returns

If we wanted to get more sophisticated in modelling share prices we might use something other than a coin toss to determine the share price return, we might draw a number from a normal distribution.

We would still have

$$S_i = (1 + R_i)S_{i-1},$$

but the R_is would now come from the distribution represented by the probability density function

$$\frac{1}{0.01708 \sqrt{2\pi}} \exp\left(-\frac{1}{2 \times 0.01708^2}(R_i - 0.001217)^2\right).$$

I've just put the measured mean and standard deviation of AAPL stock into the earlier probability density function. If you had a different stock, you'd use two different numbers.

The Distribution Of Share Prices

Suppose that the share price starts at S_0 at time zero. After one timestep we have

$$S_1 = (1 + R_1)S_0.$$

After another timestep we have

$$S_2 = (1 + R_2)S_1 = (1 + R_2)(1 + R_1)S_0.$$

And after n time steps

$$S_n = S_0 \prod_{i=1}^{n}(1 + R_i).$$

(The Π symbol means product, multiplying.) Now take logarithms of both sides and remember that the logarithm of a product is the same as the sum of logarithms,

$$\log(S_n) = \log(S_0) + \sum_{i=1}^{n} \log(1 + R_i).$$

And now for a big leap ... I won't go through the details, I'll just cut to the chase ...

A moment ago, I described the Central Limit Theorem, explaining how it says that if you add up lots of random numbers from the same distribution, no matter what that distribution is, then the result is normally distributed. Well, that's what we've got here. The R_is are random, so the $\log(1 + R_i)$s are also random, and we are adding up lots of them, so ... the logarithm of the share price in the future will be normally distributed! We say that S is lognormally distributed. (Technically we need to add up lots of random numbers, which really means a lot of small timesteps.)

Here's a plot of the lognormal distribution.

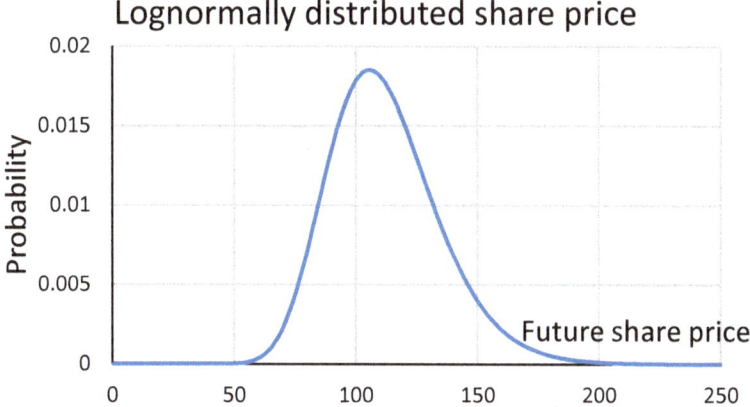

We interpret this plot this way: The future share price is most likely to be around 100, maybe a bit higher. Very unlikely to be less than about 60, but not impossible, and also unlikely to be above 180, but still possible.

The lognormal distribution is different from the normal distribution in several ways. One is that there is zero chance of the lognormal variable being zero or negative, whereas with the normal distribution all numbers, positive and negative, are possible (even if the extremes are highly unlikely).

The normal distribution is symmetrical about its mean, but not so the lognormal. The long tail of the distribution to the right means that the lognormal has what is called a positive skew.

Of course, both distributions, and all continuous distributions, have an area under the curve of one.

Aside: Simulating correlated random walks

It would be nice if we could simulate two correlated random walks by tossing two coins. But it's not quite that easy. With just four heads and tails possibilities there's not much flexibility. Instead, a good way of generating two random numbers that are correlated is to start with two *un*correlated normally distributed numbers which you then combine linearly. I'll leave that there.

Option Replication

The following table shows another way of looking at option valuation.

	Share price	Cash	Option	Hedging... 1/99 Stock - 2/99 Option	Replication... 1/2 Share - 99/2 Cash
Today	100	1	V	100/99 - 2V/99	100/2 - 99/2 = 0.5
Scenario 1	101	1	1	1	1
Scenario 2	99	1	0	1	0

The left-hand column refers to Today, Scenario 1 and Scenario 2. Today, before expiration, reading across we know the share price, it is $100, and we know what $1 is ... it's $1! But we don't know the option value, here it's called V, to be determined. But expiration could turn out in one of two ways. Read along the Scenario 1 row. In this scenario, the stock price rises to 101, $1 in the bank is still $1 if interest rates are zero. And the option payoff for a call with a strike of 100 is 1.

Reading along the Scenario 2 row, we have stock now falling to 99, cash is again $1, and the option payoff is zero.

And now we come to the two columns on the right representing *two* special portfolios.

The first special portfolio, in green, is what we've already seen. That's the hedged portfolio. The numbers are slightly different from what I used before, all I've done is multiply everything in my previous hedged portfolio by -2/99. So we have 1/99 of the stock and -2/99 of the option, with a value of $100/99 - 2V/99$ today. Now look at the value of this portfolio at expiration. In both scenarios its value is 1. This is the same as cash, in both scenarios, and so that $100/99 - 2V/99$ today must also be 1, i.e. $V = 0.5$. That was hedging.

The second portfolio, in blue, is new. All I've done is a bit of rearranging. This portfolio is half of the share and -99/2 in cash. The value of this portfolio is different in the two scenarios, in Scenario 1 it is 1 and in Scenario 2 it is zero. But those values are the same as the option payoff in the two scenarios. So its current value, 0.5, must be the same as the option value. Again, $V = 0.5$. The second special portfolio (the blue one), shows you how to *replicate* the option payoff just by using the stock and cash. You can build your own option payoff with shares and money in, or borrowed from, the bank!

EPILOGUE

So there you are, the basics of the mathematics used in high finance. A career in banking is but a degree in mathematics or computer science away!

As with any mathematical subject, there's no substitute for getting your hands dirty with actual examples. So I do insist that you try all of the exercises at the end of the last few chapters. On request, I will send all the answers to teachers. I will also send the answers to students, but only if you email me the answers to the first two questions in each section of exercises. Deal? I'm paul@wilmott.com.

Also email me if you want a few spreadsheets showing how many of the ideas are implemented in practice. Also email just for a chat, or if you find any typos!

But never forget, it's only money.

ACKNOWLEDGMENTS

Thank you to Unsplash and the following for their photographs: Elimende Inagella, Dylan Calluy, Towfiqu Barbhuiya, Martin Ceralde and Adam Nowakowski.

> 'If there is anyone to whom I owe money, I'm prepared to forget it if they are.' Errol Flynn

INDEX

Page numbers in **bold** are major entries on the subject.

'Ukulele, 124
Accountant, **25**
Actuary, **27**
Agnetha Fältskog, ix
Albert Einstein, 8
Anni-Frid Lyngstad, ix
Annual growth rate, 59
Annual Percentage Rate, **41**
Arbitrage, **21**
Arithmetic random walk, **59**
Artificial intelligence, x
At the money, **76**
Auditor, **25**
Average, **30**
Backgammon, **50**
Ballroom Dancing, 124
Bank of England, 3
Bank runs, **4**
Banks, **3**
Bartering, **1**
Base rate, **12**
Benny Andersson, ix
Bid, 21
Bid-offer spread, 21
Binomial model, **78**
Binomial trees, **90**
Björn Ulvaeus, ix
Blackjack, 10
Black-Scholes equation, 94
Black-Scholes formulae, 94
Bob Fosse, ix
Bollinger bands, 24
Bond, **13**, 37
Bond options, 47

Bubbles, **4**
Cabaret, ix
Calculus, 47, 100
Card counting, 10
Central Banks, 3
Central Limit Theorem, 114
Chess, **41**
Chief Financial Officer, **26**
Claude Shannon, 10
Columbus, 102
Commodities, **16**
Compound interest, **39**
Consumption, 73
Continuous compounding, **100**, **101**
Continuous probability distribution, **104**, 111
Continuous time, 47, 100
Conventions, **38**, 59
Correlation, 19, **34**, 60, 65
Correlation is not causation, 35
Coupon, 13
Covariance, **35**
Crashes, **4**
Credit rating, 13
Credit risk, 13
Cryptocurrency, **3**
Dab Hands, 124
Data scientist, **27**
Default, **13**
Differential equations, 100
Diffusion, 8
Discount factor, **44**
Discounting, **44**

Discrete probability distribution, **104**
Discrete time, 47
Diversification, 9, 19, **65**
Dividend, **14**
DJIA, 15
Dot-com bubble, 5
Double coincidence of wants, 2
Dow Jones Industrial Average, **6**
Ed Thorp, **10**
Efficient frontier, 9, **71**
Efficient Market Hypothesis, 8, **9**, 24
EMH, **9**
Equity, 14
Eugene Fama, **9**
European Central Bank, 3
Exchange, **14**
Exchange rate, **16**
Exercise price, 17, **76**
Exercises, **48**, **63**, **74**, **95**
Exotic options, 84
Expectation, 18, **30**
Expiration (date), 16, **76**
Expiry, **76**
Exponential function, **96**
Federal Reserve, 3
Fiat money, 2
Fibonacci series, 24
Financial Advisor, **25**
Financial Analyst, **25**
Financial Software Developer, **28**
Financial theories, **8**
Fischer Black, 10, 82
Fitch, 13
Fixed income, 15, **37**
Fixed rate, **11**, 37
Floating rate, **11**, 37, **46**
Foreign exchange, **16**

Forex, 16
Forward contract, 16
Forward rates, **45**
Fred Ebb, ix
Free lunch, 17
FTSE All World, 15
FTSE World Government Bond Index, 15
FTSE100, 15
Fundamental analysis, **23**
Futures contract, 16
FX, 16
Geometric random walk, **59**
Gold, 2
Greater Fool Theory, **5**
Growth, 18
Happy Meal, 2
Harry Markowitz, **9**
Haruspex, 102
Hedge funds, **22**
Hedging, **21**, **83**, **85**
Herding, 4
High net worth, 22
In the money, **76**
Index, 14
Insurance, 92
Interest, **11**
Investment Banker, **26**
Investment banks, 3
It's only money, 1
Jargon, **11**
Jensen's Inequality, **93**, **103**
Joel Grey, ix
John Cox, 82
John Kandor, ix
John Maynard Keynes, 23
Juggling, 124
Kelly investing, 73
Knickerbocker Trust Company, 4

Law of Large Numbers, **106**
Leverage, **22**, 76
Linear, **103**
Liza Minnelli, ix
Log tables, 99
Logarithm, **98**
Lognormal distributed, **114**
Long, 17
Long Term Capital Management, **7**, 22
Louis Bachelier, **8**
Marco Polo, 2
Mark Rubinstein, 82
Market crashes, **73**
Market sentiment, 23
Markets, **15**
Mathematics, **11**, **25**
Maturity, **12**
Mean, 30
Median, 30
Medici, 3
Members of Parliament, 36
Mode, 30
Modern Portfolio Theory, **9**, **65**
Moody's, 13
Mortgage, **12**
Myron Scholes, **7**, **10**, 82
NASDAQ Composite, 15
Natural logarithm, 98
New York Stock Exchange, 6
Nikkei 225, 15
No such thing as a free lunch, 17
No-arbitrage, **82**, **83**, **93**
Nobel Prize in Economics, 3, 7, 9, 10, 61, 75, 80
Normal distribution, **110**
October 20th, 1987, **5**
Offer, 21
Options, **10**, **16**, **75**

Ordinary differential equations, 47
Out of the money, **76**
Panic of 1907, 4
Paper money, **2**
Partial differential equation, 94
Partial differential equations, 62
Pascal's Triangle, **107**
Paul Wilmott, iv, **124**
Payoff, **76**
Payoff diagrams, **77**
Portfolio, 19, **65**
Portfolio Manager, **27**
Premium, 76, 92
Present value, **44**
Principal, **12**
Probabilistic model, 24
Probability density function, **105**, **110**, 111
Put options, **91**
Put-call parity, **93**
Quant, **29**
Quantitative analysis, **24**
Quantitative Analyst, **29**
Quantitative Developer, **28**
Random walk, **58**
Range, 32
Real probabilities, **87**
Replication, **117**
Reserves, 12
Retail bank, 3
Return, **18**
Risk, **18**
Risk averse, **20**, 67
Risk Manager, **28**
Risk neutral, **20**
Risk neutrality, 20
Risk seeking, **20**
Risk-neutral probabilities, **87**
Risk-neutral world, **87**

Robert C. Merton, **7**, 10, 82
Roulette, 10
Royal Statistical Society, 36
S&P GSCI, 15
S&P500, 15
Scientific calculator, 97, 98
Securities Trader, **26**
Share, **14**
Sharpe Ratio, **61**, **73**
Shells, 2
Short, 17
Silver, 2
Simple interest, **39**
Simulations, **55**
Skew, 115
Slide rule, 99
Standard deviation, 18, **32**
Standard normal distribution, **110**, **111**
Stanton Road County Primary School, 124
Statistical arbitrage, 21
Stephen Hawking, 1
Steve Ross, 82
Stochastic calculus, 8, 62, 82, 94
Stockbroker, **26**
Stress test, 28
Strike price, 17, **76**
Supply and demand, 21
Swaps, 47

Tangency Portfolio, **73**
Taylor series, 97, 100
Technical analysis, **24**
Tends to infinity, 100
The Central Limit Theorem, **107**
There's no such thing as a free lunch, **21**, **82**
Tilde, 101
Time value of money, 44
Trees, **90**
Triangular arbitrage, **21**
Trick question, 80
Tulip Mania, **5**
Twiddles, 101
Two wrongs do make a right, 89
Underlying (asset), **76**
US Dollar Index, 15
Utility functions, 73
Value at risk, 28
Variance, **32**
Venture Capitalist, **26**
Volatility, 19, **60**, 89
Wearable computer, 10
William Sharpe, 61
Wirral Grammar School for Boys, 124
μ, 60
σ, 60

ABOUT PAUL WILMOTT

Paul is a practising mathematician. He researches in mathematics, teaches mathematics, and writes about mathematics.

His love of mathematics started at Stanton Road County Primary School in Bebington, Merseyside. He fondly remembers his first ever homework, pages and pages of manipulations of fractions. Wirral Grammar School for Boys followed, where he took as many mathematical subjects as possible. He then studied mathematics at St Catherine's College, Oxford, where he also received his doctorate. In mathematics.

Paul was briefly a professional juggler with the Dab Hands troupe, and was an undercover investigator for the UK's Channel 4 TV. He also has three half blues from Oxford University for Ballroom Dancing. And he plays the 'ukulele.

> 'A father is someone who carries pictures of his children in his wallet where his money used to be.' Anonymous

ALSO BY PAUL WILMOTT

The Mathematics of Artificial Intelligence for High Schoolers

This is the book that introduces schoolchildren to the mathematics behind Artificial Intelligence!

There are many books with no mathematics, just the description of what AI can do. Then there are many books with no mathematics, just lots of computer code. And there are also quite a few books with lots and lots of mathematics but which are incomprehensible without a degree in mathematics.

However there aren't books like this one!

This book is for high-school children who are confident in their mathematics. They do not have to be advanced or even top of their class. But they do have to be comfortable with learning new concepts.

The book starts with a chapter on the history of AI. This is followed by a chapter of jargon, useful mathematics, and a description of the main techniques. Then it's one chapter on each of Nearest Neighbours, Regression, Clustering, Decision Trees, Neural Networks and Reinforcement Learning. Each of these in-depth chapters covers uses of the technique, the mathematical details, a real example and then a project for readers to work through to work through on their own or in teams. The book ends with sources for some useful data.

www.ingramcontent.com/pod-product-compliance
Lightning Source LLC
Chambersburg PA
CBHW041318110526
44591CB00021B/2824